Aktuelle Forschung Medizintechnik

Editor-in-Chief:
Th. M. Buzug, Lübeck, Deutschland

Unter den Zukunftstechnologien mit hohem Innovationspotenzial ist die Medizintechnik in Wissenschaft und Wirtschaft hervorragend aufgestellt, erzielt überdurchschnittliche Wachstumsraten und gilt als krisensichere Branche. Wesentliche Trends der Medizintechnik sind die Computerisierung, Miniaturisierung und Molekularisierung. Die Computerisierung stellt beispielsweise die Grundlage für die medizinische Bildgebung, Bildverarbeitung und bildgeführte Chirurgie dar. Die Miniaturisierung spielt bei intelligenten Implantaten, der minimalinvasiven Chirurgie, aber auch bei der Entwicklung von neuen nanostrukturierten Materialien eine wichtige Rolle in der Medizin. Die Molekularisierung ist unter anderem in der regenerativen Medizin, aber auch im Rahmen der sogenannten molekularen Bildgebung ein entscheidender Aspekt. Disziplinen übergreifend sind daher Querschnittstechnologien wie die Nano- und Mikrosystemtechnik, optische Technologien und Softwaresysteme von großem Interesse.

Diese Schriftenreihe für herausragende Dissertationen und Habilitationsschriften aus dem Themengebiet Medizintechnik spannt den Bogen vom Klinikingenieurwesen und der Medizinischen Informatik bis hin zur Medizinischen Physik, Biomedizintechnik und Medizinischen Ingenieurwissenschaft.

Editor-in-Chief:
Prof. Dr. Thorsten M. Buzug
Institut für Medizintechnik,
Universität zu Lübeck

Editorial Board:
Prof. Dr. Olaf Dössel
Institut für Biomedizinische Technik,
Karlsruhe Institute for Technology

Prof. Dr. Heinz Handels
Institut für Medizinische Informatik,
Universität zu Lübeck

Prof. Dr.-Ing. Joachim Hornegger
Lehrstuhl für Bildverarbeitung,
Universität Erlangen-Nürnberg

Prof. Dr. Marc Kachelrieß
German Cancer Research Center,
Heidelberg

Prof. Dr. Edmund Koch,
Klinisches Sensoring und Monitoring,
TU Dresden

Prof. Dr.-Ing. Tim C. Lüth
Micro Technology
and Medical Device Technology,
TU München

Prof. Dr. Dietrich Paulus
Institut für Computervisualistik,
Universität Koblenz-Landau

Prof. Dr. Bernhard Preim
Institut für Simulation und Graphik,
Universität Magdeburg

Prof. Dr.-Ing. Georg Schmitz
Lehrstuhl für Medizintechnik,
Universität Bochum

Yulia Levakhina

Three-Dimensional Digital Tomosynthesis

Iterative Reconstruction, Artifact Reduction and Alternative Acquisition Geometry

Yulia Levakhina
University of Lübeck
Germany

Dissertation University of Lübeck, 2013

ISBN 978-3-658-05696-4   ISBN 978-3-658-05697-1 (Ebook)
DOI 10.1007/978-3-658-05697-1

The Deutsche Nationalbibliothek lists this publication in the Deutsche Nationalbibliografie; detailed bibliographic data are available in the Internet at http://dnb.d-nb.de.

Library of Congress Control Number: 2014937470

Springer Vieweg
© Springer Fachmedien Wiesbaden 2014
This work is subject to copyright. All rights are reserved by the Publisher, whether the whole or part of the material is concerned, specifically the rights of translation, reprinting, reuse of illustrations, recitation, broadcasting, reproduction on microfilms or in any other physical way, and transmission or information storage and retrieval, electronic adaptation, computer software, or by similar or dissimilar methodology now known or hereafter developed. Exempted from this legal reservation are brief excerpts in connection with reviews or scholarly analysis or material supplied specifically for the purpose of being entered and executed on a computer system, for exclusive use by the purchaser of the work. Duplication of this publication or parts thereof is permitted only under the provisions of the Copyright Law of the Publisher's location, in its current version, and permission for use must always be obtained from Springer. Permissions for use may be obtained through RightsLink at the Copyright Clearance Center. Violations are liable to prosecution under the respective Copyright Law.
The use of general descriptive names, registered names, trademarks, service marks, etc. in this publication does not imply, even in the absence of a specific statement, that such names are exempt from the relevant protective laws and regulations and therefore free for general use.
While the advice and information in this book are believed to be true and accurate at the date of publication, neither the authors nor the editors nor the publisher can accept any legal responsibility for any errors or omissions that may be made. The publisher makes no warranty, express or implied, with respect to the material contained herein.

Printed on acid-free paper

Springer Vieweg is a brand of Springer DE.
Springer DE is part of Springer Science+Business Media.
www.springer-vieweg.de

Preface by the Series Editor

The book *Three-Dimensional Digital Tomosynthesis: Iterative Reconstruction, Artifact Reduction and Alternative Acquisition Geometry* by Dr. Yulia M. Levakhina is the 14th volume of the Springer-Vieweg series of excellent theses in medical engineering. The thesis of Dr. Levakhina has been selected by an editorial board of highly recognized scientists working in that field.

The Springer-Vieweg series aims to establish a forum for Monographs and Proceedings on Medical Engineering. The series publishes works that give insights into the novel developments in that field.

Prospective authors may contact the Series Editor about future publications within the series at:

Prof. Dr. Thorsten M. Buzug
Series Editor Medical Engineering
Institute of Medical Engineering
University of Lübeck
Ratzeburger Allee 160
23562 Lübeck
Web: www.imt.uni-luebeck.de
Email: buzug@imt.uni-luebeck.de

Foreword

Conventional X-ray imaging suffers from the drawback that it only produces two-dimensional projections of a three-dimensional object. This results in a reduction in spatial information (although an experienced radiologist might be able to compensate for this). In any case, a projection represents an averaging. The result of the averaging can be imagined if one were to overlay several radiographic sections at the light box for diagnosis. It would be difficult for even an expert to interpret the results, as averaging comes along with a considerable reduction in contrast, compared with the contrast present in one slice.

In the 1920s, the desire to undo the averaging process that characterizes conventional X-ray radiography led to the first tomographic concept. The word tomography was considerably influenced by the Berlin physician Grossmann, whose Grossmann tomograph was able to image one single slice of the body. The principle of the conventional or analog geometric tomography method is very simple. During image acquisition, the X-ray tube is linearly moved in one direction, while the X-ray film is synchronously moved in the opposite direction. For this reason, only points in the plane of the rotation center are imaged sharply. All points above and below this region are blurred, more so at greater distances from the center of rotation. However, blurred information above and below the center of rotation does not disappear, but is superimposed on the sharp image as a kind of gray veil or haze. Therefore, a substantial reduction in contrast is noticeable.

This book on *Three-Dimensional Digital Tomosynthesis: Iterative Reconstruction, Artifact Reduction and Alternative Acquisition Geometry* summarizes the research work of Dr. Yulia Levakhina. The work has been carried out at the Institute of Medical Engineering at the University of Lübeck. It focuses on image-improvement methods for a tomosynthesis device working with insufficient and inconsistent projection data.

If the raw projection data to be used for 3D reconstruction in X-ray imaging are insufficient and/or inconsistent, artifacts cover the reconstructed objects that reduce the diagnostic value of the images significantly. However, digital tomosynthesis is a concept that is based on the reconstruction of three-dimensional volumes from a few projections.

This book concludes the results of a number of original papers and innovations Dr. Yulia Levakhina has achieved in the discipline of digital tomosynthesis. A new method for the reduction of out-of-focus artifacts and an innovative acquisition geometry are spotlights that significantly exceed the current state-of-the-art.

November 2013, Lübeck

Prof. Dr. Thorsten M. Buzug
Institute of Medical Engineering
University of Lübeck

It has been said that something as small as the flutter of a butterfly's wing can ultimately cause a typhoon halfway around the world - Chaos Theory
"The Butterfly Effect" (film)

Acknowledgments

When four and a half years ago I arrived from Moscow to Lübeck with a suitcase with some clothes and a "fresh" diploma, I could have not even imagined what wonderful and challenging journey is awaiting for me. During this time I have collected a lot of stuff which now can fill a half of an LKW-car and I have gained a valuable experience which is unmeasurable. But what is the most important, is that I met a lot of incredible people without whom I would probably have never reached my goals.

I am very lucky person, because I had such a great supervisor as Prof. Thorsten M. Buzug. I thank him for his help throughout the years in moments when I needed help. A the same time he gave me a chance to try everything on my own and to find my own way: it was hard but very important on the long run. Thank him, I learned how to do research and how *to think*. This helped me a lot to grow professionally and personally. While being quite busy person, he always found a time for a scientific discussion or for proofreading of a paper even if it was finished a few hours before the deadline. Yes, *every* student including myself start to write at the last moment and I am thankful for his understanding. He believed that I can find a solution to any problem and that I can accomplish any task. Indeed, with his support, optimism and enthusiasm, each time I succeed.

I must say a special thank to you the Graduate School for Computing in Medicine and Life Sciences and DFG for funding this project and for my scholarship. And again thanks Prof. Buzug for admitting me to this program. Based on my motivation letter and on the list of my grades he invited me to a personal interview (from Moscow) and organized a very exciting day at the Institute of Medical Engineering (IMT) with a pizza in the evening. Few weeks after the interview I got an acceptance and with no doubts started to pack my luggage. The last six monthes of my study were supported by the company YXLON International GmbH, Hamburg.

Acknowledgments

During my PhD time I traveled as much as I never done in my life before. I had an opportunity to visit a number of conferences and summer schools and to presented my research to the CT community. In such events I meet many interesting people (including Prof. J. Fessler, Prof. E. Todd Quinto and Dr. B. de Man), I learned a lot from their talks and I was inspired by exchanging ideas.

But of course, the main refreshing insight and idea generation came from my colleagues at IMT (listed in a random order): Alex Opp, Andreas Mang, Sven Biederer, Kerstin Ludtke-Buzug, Matthias Kleine, Maik Stille, Gael Bringout, Matthias Graeser, Alina Toma, Tina Anne Schuetz, Stefan Becker, Maren Bobek, Mandy Grüttner, Tomas Weidinger. Thank you for all nice moments (also listed in a random order): discussing, talking, chatting (also at nights), drinking coffee, reading papers, doing science, learning, teaching, going out, shopping, proofreading, writing, solving IT-problems, going to partys, doing sport, having fun at conferences. I had a great time with my office-mate Bärbel Kratz who shared an office with me. Many thanks goes also to all undergraduate students at IMT and especially to my bachelor student Aileen Cordes and my master student Sylvia Kiencke. While supervising your thesis I learned quite a lot of new things. I wish both of you great success with your PhD study.

I shall thank my colleges from Radiology department for cooperation and expertise in medicine: Prof J. Barkhausen, Dr. F. Vogt and Robert Duschka. I owe thousand thanks to Robert for doing measurements with me, sometimes staying at work late or even till midnight.

Without Sergey Tarassenko I would have never ended up with an idea of doing PhD in Germany. Thank you for this and for your support during the first two years of my study.

This list is definitely not complete without my friends from Moscow and my family. Thanks for not forgetting me and always looking forward when I am coming back for a vacation. Thanks to all of you for remaining my friends even if there are almost 2000 km between us, thanks for all the support you gave me, thanks for believing in me. I miss all of you: Tatjana and Max Kurganskiy, Mischa Kruchkov, Konstantin Barsukov, Alex Asmodeev, Larisa Levakhina and Konstantin Mozgov.

Last but not least, my deep appropriation goes to Jan Müller. Thank you for being my science-mate and soul-mate, for your your love, support and unwavering belief in me.

November 2013, Lübeck   Yulia Levakhina

Abstract

Digital tomosynthesis (DT) is an X-ray based limited angle imaging technique. It is a non-invasive and non-destructive method for three-dimensional visualization of the inner structures of an object. Tomosynthesis is historically the first X-ray based tomographic technique. However, it has been forgotten with the development of computed tomography (CT). Only recently, developments in the field of digital X-ray detectors and computer technologies have led to a renewed interest in this technique. A high in-plane resolution, three-dimensionality and a low radiation dose make DT an attractive alternative to CT in many imaging applications. The most widely used DT application in medical imaging is breast imaging. In this thesis an alternative application of tomosynthesis for imaging of hands is considered.

In contrast to CT, the DT projection dataset is incomplete, because the X-ray source and the detector do not completely rotate around the patient. The incompleteness of the dataset violates the tomographic sufficiency conditions and results in limited angle artifacts in the reconstructed images. Although DT is a volumetric imaging technique and provides dimensional information about the location of structures, the complete three-dimensional information about the object cannot be reconstructed. Therefore, one of the major issues is the improvement of the tomosynthesis image quality.

This thesis addresses the connection of the reconstruction problem and the incompleteness of the DT dataset. The main aim is to understand the factors, which cause the formation of limited angle artifacts and, thus, to account for them in order to improve the image quality and the axial resolution.

A thorough literature review on the tomosynthesis topic is presented in each chapter. A three-dimensional tomosynthesis reconstruction framework including fast and accurate forward- and backprojectors and flexible geometry, has been developed to study several aspects of DT. All experimental studies presented in this thesis use simulated data and real clinical data of hands.

Two conceptually different strategies for improving the image quality are investigated. The first strategy deals with reconstruction algorithms. Within this strategy a non-linear backprojection is used in the simultaneous algebraic reconstruction technique (SART). The non-linear backprojection is based on a spatially-adaptive weighting scheme which is designed to reduce out-of-focus artifacts caused by high-absorption structures. The novel concepts of the backprojected space representation and a dissimilarity degree are proposed to construct this weighting scheme. It will be shown that the weighted SART reduces contribution of high-absorption structures to the formation of artifacts on out-of-focus slices while preserving these features in the in-focus slices.

The second strategy is based on the assumption that the incompleteness degree of the dataset can be reduced by using more appropriate acquisition geometry. The impact of several acquisition parameters to the tomosynthesis image quality for the standard geometry used in clinics is studied. In the presented study the limitations of the standard geometry will be demonstrated. Although the image quality can be improved by acquiring data over a wider angular range, above a certain threshold this becomes infeasible. Therefore, the results motivate to search for an alternative acquisition geometry. A novel dual-axis acquisition geometry with a tiltable platform will be proposed. The data in the original direction are acquired using the X-ray tube movement and the data in the additional perpendicular direction are acquired by tilting the object. The projection data acquired along two axes have less incompleteness. Based on a simulation study it will be shown that such acquisition geometry results in less artifacts and improves the axial resolution.

The findings and conclusions of this work have a number of important implications for future research, therefore, the suggestions for further work are given for each addressed topic.

Contents

1 **Introduction**   1
    1.1 X-ray imaging and tomosynthesis 1
    1.2 Contribution of this work 2
    1.3 Outline of the thesis 5
    1.4 Publications 6

2 **Where we are today: Tomosynthesis research and development**   11
    2.1 Tomosynthesis basics 12
       2.1.1 Introduction 12
       2.1.2 Tomosynthesis technology 16
       2.1.3 DICOM format 17
    2.2 From radiostereoscopy to digital tomosynthesis 18
    2.3 Evolution of reconstruction algorithms 23
    2.4 CT and tomosynthesis today: practical comparison 26
       2.4.1 Technical parameters 26
       2.4.2 Reconstructed images 27
    2.5 Problems of limited data tomography 33
       2.5.1 Types of limited data problems 33
       2.5.2 Radon transform and singularities 34
       2.5.3 Incomplete Fourier space 36
       2.5.4 Tuy-Smith sufficiency condition 38
       2.5.5 Artifacts and limited resolution 38

3 Forward and backprojection model (FP/BP) — 43
- 3.1 Discrete image representation using series expansion — 44
- 3.2 Pixel basis functions — 45
- 3.3 FP/BP algorithms for pixel basis functions — 46
- 3.4 Kaiser-Bessel basis functions (blobs) — 49
  - 3.4.1 Properties of the blobs — 49
  - 3.4.2 Finding the optimal parameters a, α, m — 53
- 3.5 FP/BP algorithms for blob basis functions — 55
  - 3.5.1 Ray tracing through grid of blobs — 55
  - 3.5.2 Lookup table calculation — 57
- 3.6 Efficient distance-driven projector in 2D — 58
  - 3.6.1 General strategy — 58
  - 3.6.2 Unification of four angular cases — 60
  - 3.6.3 The sweep line principle for pixels — 63
  - 3.6.4 The sweep line principle for blobs — 67
- 3.7 Efficient distance-driven projector in 3D — 69
  - 3.7.1 General strategy — 69
  - 3.7.2 Unification of the angular cases — 71
  - 3.7.3 Sweep line in three dimensions — 72

4 Iterative image reconstruction for tomosynthesis — 75
- 4.1 Discrete model of the physical system — 76
- 4.2 Iterative reconstruction schemes — 77
  - 4.2.1 Algebraic reconstruction — 78
  - 4.2.2 Statistical Reconstruction — 79
- 4.3 Considerations for practical implementation of SART — 82
  - 4.3.1 How to address tomosynthesis datasets: a dictionary approach — 82
  - 4.3.2 Memory handling in MATLAB®: the base workspace — 83
  - 4.3.3 Memory costs — 84
  - 4.3.4 Computational complexity — 86
- 4.4 Projection access order for SART — 87
  - 4.4.1 Literature review — 87
  - 4.4.2 Sequential order — 88
  - 4.4.3 Random permutation — 88
  - 4.4.4 Golden ratio — 89
  - 4.4.5 Prime numbers decomposition (PND) — 89
  - 4.4.6 Prime number increment (PNI) — 93

Contents

	4.4.7 Multilevel scheme (MLS)	93
	4.4.8 Weighted distance scheme (WDS)	94
	4.4.9 Data-based minimum total correlation order	94
	4.4.10 Simulation results	95

5 Backprojected space in image reconstruction — 99

- 5.1 Theory of backprojected space … 100
  - 5.1.1 Stackgram representation in literature … 100
  - 5.1.2 Properties of BP-space … 101
- 5.2 A weighting scheme based on dissimilarity … 106
  - 5.2.1 Motivation for weighting: tomosynthesis blur formation … 106
  - 5.2.2 Dissimilarity degree … 108
  - 5.2.3 Weighting scheme … 109
- 5.3 Non-linear backprojection ωBP for tomosynthesis … 111
  - 5.3.1 Introducing weighting in the BP operator … 111
  - 5.3.2 Demonstration of the dissimilarity and weighting … 114
  - 5.3.3 Reconstruction results … 117
- 5.4 Weighted ωSART for tomosynthesis … 120
  - 5.4.1 Introducing the non-linear BP into SART … 120
  - 5.4.2 Computational complexity and implementation strategy … 120
  - 5.4.3 Reconstruction results … 122
- 5.5 Weighted ωSART for metal artifact reduction in CT … 125
  - 5.5.1 Parameter γ … 125
  - 5.5.2 Reconstruction results … 128
- 5.6 Interpolation in BP-space for metal artifact reduction in CT … 133
  - 5.6.1 θ-interpolation in BP-space … 133
  - 5.6.2 Preliminary results … 135

6 Dual-axis tilt acquisition geometry — 137

- 6.1 Tomosynthesis "mini" simulator … 138
  - 6.1.1 A finger bone software phantom … 138
  - 6.1.2 Simulation software … 139
  - 6.1.3 Image quality metrics for performance evaluation … 141
- 6.2 Influence of the system acquisition parameters … 142
  - 6.2.1 The impact of the angular range θ … 142
  - 6.2.2 Influence of the number of projections N_{proj} … 146
  - 6.2.3 Influence of the angular step size $\Delta\theta$ … 149

    6.3 A novel geometry: hybrid dual-axis tilt acquisition 151

        6.3.1 How to acquire more information of the object? 151

        6.3.2 Theoretical background 153

        6.3.3 Singularities of Radon transform and limited data 154

        6.3.4 Incomplete Fourier space 154

        6.3.5 Tuy-Smith sufficiency condition 156

        6.3.6 Angle in x-direction and re-distribution of projections. 157

        6.3.7 Influence of number of projections 159

    6.4 Influence of the object orientation 161

7 Conclusions and suggestions for further work   163

8 Appendix: MATLAB®   167

    8.1 MATLAB® File Exchange 167

    8.2 PubMed trend search 168

9 References   169

Chapter 1

Introduction

Contents

1.1	X-ray imaging and tomosynthesis	1
1.2	Contribution of this work	2
1.3	Outline of the thesis	5
1.4	Publications	6

1.1 X-ray imaging and tomosynthesis

X-ray based imaging techniques are already the subject of active research for almost 130 years starting right after the discovery of X-rays by Wilhelm Conrad Röntgen in 1895 (Roentgen 1895a, Roentgen 1895b) up to the present. The nature of X-rays to penetrate the object which has been used to visualize the inner structures of opaque objects have changed the world in medical and non-medical application fields. It opened new opportunities in recognition and understanding of human diseases without surgical intervention and in non-destructive material testing (NDT). Although the human body is "transparent" for X-rays and can be visualized using X-rays, a single X-ray image contains only the projective overlap of all structures in the body. The three-dimensional information about the structure locations can be recovered only by using the principle of tomography. For this, a set of X-ray images from different sides must be measured and the inverse problem of image reconstruction must be solved.

Historically, the first X-ray imaging modality which aims to visualize an object in three-dimensions was tomosynthesis. In tomosynthesis projection images are acquired over a limited angular range. In general, this is not enough to reconstruct the object exactly. However, some three-dimensional information can still be recovered but the image quality is degraded by blurring out-of-focus artifacts. Back in the 1930s to 1970s, tomosynthesis was a promising imaging modality and a lot of effort was given to improve its performance in terms of speed and to obtain images with less artifacts. A lack of digital X-ray detector technology was the main stopping factor in the development of tomosynthesis. With the development of the true tomographic principle in 1972 (Hounsfield 1973, Ambrose 1973), in which the data is obtained over the 360° angular range, tomosynthesis was abandoned because of the clear advantages of CT to produce slices of an object without typical tomosynthesis blurring artifacts. Tomosynthesis has regained scientific interest in the beginning of the 21st century because of technological advances. The combination of fast digital flat-panel X-ray detectors and improved computer technologies offered a solution to the problem of long examination time and long processing time. It made tomosynthesis practically feasible. Nowadays, tomosynthesis is one of the "hot topics" in the field of X-ray based tomographic imaging (Sechopoulos 2013a, Sechopoulos 2013b). The main field of tomosynthesis application is breast imaging. Alternative applications also exist. The focus of this thesis is tomosynthesis with application to the imaging of human hands. Tomosynthesis is an attractive alternative to CT and computed radiography (CR) for imaging of hands because DT combines the simplicity, high resolution and low dose of CR and the three-dimensionality of CT.

1.2 Contribution of this work

With the development of advanced detectors and PC, digital tomosynthesis is again of great interest among scientists. However, the problem of data incompleteness of the projection dataset does not disappear. The incompleteness of the data violates tomographic sufficiency conditions and results in images with artifacts and limits the in-depth resolution. This makes an accurate image reconstruction a very challenging task.

The main goals of this work are to understand what influences the tomosynthesis performance in terms of image quality and artifacts and to propose methods to improve the tomosynthesis performance. An understanding of the tomosynthesis topic in general is important, therefore an intensive literature review on tomosynthesis history, existing methodology, the state of the art and open problems will be presented. Additionally, the study of the related CT subjects and the adaptation of several CT algorithms for tomosynthesis will be given.

1.2 Contribution of this work

Two different approaches to improve tomosynthesis performance will be proposed in this work. The first approach is based on the optimization of the reconstruction strategy for the given limited data. Given the measured tomosynthesis data, a suitable reconstruction algorithm is required to provide images with less artifacts and better quality. This includes the choice of the reconstruction algorithm and its parameters as well as an accurate implementation. The second approach is based on the acquisition of more reliable data using an adapted acquisition geometry. It can improve the image quality and resolution because the acquisition parameters and geometry influence the incompleteness degree of the obtained data. If the data incompleteness is reduced, the image quality and resolution will be improved. As such geometry, a novel dual-axis acquisition geometry will be proposed.

The contributions of this work are following

- **A topical review** which includes a thorough literature review on tomosynthesis and a comparison of the state of the art tomosynthesis device with CT and micro-CT devices. Since tomosynthesis and CT are closely related, also the review of CT literature is necessary for several topics. The obtained knowledge, then, is adapted and applied for tomosynthesis.
- **A short summary of limited angle tomography** which explains where the limited angle artifacts come from.
- **Implementation the reconstruction toolbox** for three-dimensional tomosynthesis from the scratch using the knowledge from CT. Each chapter of this thesis (if applicable) contains corresponding consideration regarding practical implementation of algorithms in MATLAB® and C++ (mex). The toolbox includes
  - fast and accurate forward- and backprojector (FB and BP) for two- and three-dimensions [2 - 4], [15];
  - several standard iterative algebraic (SART) and statistical reconstruction algorithms with the possibility to vary parameters (number of iterations, initial guess, projection access order) [1], [4], [5];
  - a weighted version of simple backprojection (ωBP) and a weighted algebraic reconstruction (ωSART) with an adaptive weighting scheme and a flexible control of the weighting parameters for tomosynthesis and CT [6 - 8], [11];
  - a flexible geometry with the possibility to change the acquisition parameters (number of projections, the angular range and the angular step size), distances (source-to-object, source-to-detector, source-to-isocenter) and the X-ray tube trajectory [9], [10];

- methods to construct the backprojected space representation in two- and three-dimensions [6 - 8], [11].

- **Finding an optimal implementation** of every single component of the reconstruction toolbox. It includes
  - a study of the advantages and drawbacks, accuracy, complexity and possibility of fast implementation of FB and BP methods for CT [3];
  - fast and accurate implementation of the distance-driven projector for CT and tomosynthesis with only one loop for all angular cases [3], [4];
  - a method for an optimal memory handling for processing large tomosynthesis datasets [3], [4];
  - a method to address a large number of variables from different datasets (reconstruction and projections) [3], [4].

- **A parameter optimization analysis** which includes
  - a discussion of basis functions for image representation and interpolation strategy for FP and BP [2], [3];
  - a parameter optimization for SART on the example of the projection access order [5], [14];
  - a study on parameters for the dissimilarity-based weighting scheme for tomosynthesis and additionally for metal artifact reduction in CT;
  - a study in the impact of the geometry acquisition parameters on tomosynthesis performance [10], [12], [13].

- **Novel ideas** which include
  - a usage of the distance-driven FP and BP algorithm for tomosynthesis and benefit from the fixed detector geometry [3], [4];
  - a novel data-based projection access order for SART based on the minimum correlation [5], [14];
  - the backprojected space representation as a generalization of the stackgram approach [6 - 8], [11];
  - data dissimilarity coefficients in BP-space [6 - 8], [11];
  - a weighing scheme for tomosynthesis based on the dissimilarity in BP space for simple backprojection and for SART [6 - 8], [11];
  - BP-space for metal artifact reduction in CT;
  - a novel dual-axis acquisition geometry for tomosynthesis [9], [10].

1.3 Outline of the thesis

This doctoral dissertation consists of seven chapters and an appendix. This chapter is an introduction to this work. The next chapter gives an extensive overview on the tomosynthesis topic, including the historical development of tomosynthesis technology and its reconstruction algorithms. A state of the art tomosynthesis device will be compared with the CT and micro-CT devices in terms of technical parameters and the obtained images. Theoretical aspects of image reconstruction from limited data will be given, with focus on the explanation of why limited data is not enough and to which problems it leads.

The problem of efficient forward and backprojection algorithms will be addressed in chapter 3. A discrete image implementation using series expansion and the choice of the basis functions will be discussed. The literature review on the forward- and backprojector will be given for two common choices of basis functions: square pixels and spherically symmetric Kaiser-Bessel functions (blobs). The state of the art distance-driven algorithm will be discussed in detail and an efficient implementation for both, the pixels and the blobs basis functions for two-dimensional fan beam CT will be proposed. The discussion of the implementation strategy will be extended to the three-dimensional cone-beam CT and adapted for tomosynthesis geometry with a fixed detector. This way, FP and BP, which are necessary for iterative reconstruction, can be implemented accurate and fast in two- and three-dimensions.

The formulation of the reconstruction problem as an optimization problem, leading to iterative schemes will be given in chapter 4. Two types of iterative reconstruction will be mentioned, namely algebraic and statistical reconstruction. The focus of the rest of the thesis will be on the SART algorithm. An efficient implementation strategies of SART for three-dimensional tomosynthesis will be discussed with respect to the problem of data handling, memory and computational costs. The problem of projection access order for SART will be discussed, including a review of existing methods which can be found in CT literature and adaptation of them for tomosynthesis. A novel data-based minimum correlation approach which uses the object-related information will be proposed. The methods will be compared using a simulation study with the application to tomosynthesis.

In chapter 5, a backprojected space representation will be proposed. Its main application is to construct a data-based weighting scheme, which will be included in the backprojection operator for tomosynthesis. The resulting non-linear backprojection algorithm is designed to reduce tomosynthesis artifacts from high attenuating structures. Moreover, the ωBP can be included into the SART reconstruction to use the weighting scheme more efficiently. The second presented application of the BP-space is metal artifact reduction in CT. The weighted scheme proposed for tomosynthesis will be extended to the 360 degree CT data. It will be shown that metal artifacts can be reduced if the weighting parameters are chosen properly. Additionally, the BP-representation offers an easy method to follow the sinogram flow, which, as it will be shown, can be used for sinogram interpolation. Experiments are based on real tomosynthesis and CT data.

In addition to the reconstruction strategy to reduce artifacts, one can try to obtain more meaningful data, taking into account that the acquisition parameters must be also optimized. In chapter 6 the tomosynthesis geometry will be analyzed with respect to the improvement of tomosynthesis performance. First, the impact of the acquisition parameters will be studied and the state of the art findings will be compared with the simulation results for tomosynthesis imaging of hands. The main role of chapter 6 is to introduce an alternative acquisition geometry using a tiltable platform. In this geometry, the projection data are acquired along two axes instead of one axis. This way, the degree of data incompleteness and therefore the artifacts can be reduced and the axial resolution can be improved.

The last chapter summarizes the conclusions and gives suggestions for further work for each addressed topic.

All presented results are based on software phantom simulations and real measured data[1]. The simulation and reconstruction framework is written in Matlab/C++.

1.4 Publications

The presented work resulted in several publications as a first author which include a number of conference contributions [1 - 5] and [7 - 9] as well as two journal papers [10, 11] and a patent application [6]. Other work has been published in cooperation with other colleagues as a co-author [12 - 20]. For each publication the page number is given (if applicable) where it has been cited in this dissertation.

First author

[1] Y. M. Levakhina and T. M. Buzug. Algebraic reconstruction versus statistical reconstruction methods in CT. In *World Congress on Medical Physics and Biomed-*

---
[1] The human cadavers - respectively bodies/heads/arms/legs feet etc. as parts of cadavers - were used and dissected in this examination under permission of the 'Gesetz über das Leichen-, Bestattungs- und Friedhofswesen (Bestattungsgesetz) des Landes Schleswig-Holstein vom 04.02.2005, Abschnitt II, ğ 9 (Leichenöffnung, anatomisch)'. In this case it is allowed to dissect the bodies of the donators (Körperspender/in) for scientific and/or educational purposes.

1.4 Publications

ical Engineering, Springer IFMBE Series, volume 25, Post-Deadline Poster 37, Munich, Germany, September 2009.

[2] Y. M. Levakhina, B. Kratz, and T. M. Buzug. Two-step metal artifact reduction using 2D-NFFT and spherically symmetric basis functions. In *Nuclear Science Symposium Conference Record, IEEE*, pages 3343–3345, 2010.

[3] Y. M. Levakhina and T. M. Buzug. Distance driven projection and backprojection for spherically symmetric basis functions. In *Nuclear Science Symposium Conference Record, IEEE*, pages 2894–2897, 2010. 56

[4] Y. M. Levakhina, R. L. Duschka, J. Barkhausen, and T. M. Buzug. Digital tomosynthesis of hands using simultaneous algebraic reconstruction technique with distance-driven projector. In *11th International Meeting on Fully Three-Dimensional Image Reconstruction in Radiology and Nuclear Medicine*, pages 167–170, 2011. 25, 140

[5] Y. M. Levakhina, B. Kratz, R. L. Duschka, F. Vogt, J. Barkhausen, and T. M. Buzug. Reconstruction for musculoskeletal tomosynthesis: a comparative study using image quality assessment in image and projection domain. In *Nuclear Science Symposium Conference Record, IEEE*, pages 2569–2571, 2011. 25

[6] Y. M. Levakhina and T. M. Buzug. Verfahren zur verbesserten Vermeidung von Artefakten bei der digitalen Tomosynthese mit iterativen Algorithmen, 2011. DE patent 10 2011 115 577.9.

[7] Y. M. Levakhina, J. Mueller, R. L. Duschka, F. M. Vogt, J. Barkhausen, and T. M. Buzug. Algebraic tomosynthesis reconstruction with spatially adaptive updating term. In *Proceedings of The Second International Conference on Image Formation in X-Ray Computed Tomography*, pages 46–49, 2012. 100, 106

[8] Y. M. Levakhina, R. L. Duschka, F. M. Vogt, J. Barkhausen, and T. M. Buzug. An adaptive spatially-dependent weighting scheme for tomosynthesis reconstruction. In *Proc. SPIE*, volume 8313, pages 831350–831356, 2012. 24, 100, 106, 111

[9] Y. M. Levakhina, R. L. Duschka, F. M. Vogt, J. Barkhausen, and T. M. Buzug. A novel acquisition scheme for higher axial resolution and improved image quality in digital tomosynthesis. In *Biomed Tech*, volume 57, pages 111–114, 2012. 137

[10] Y. M. Levakhina, R. L. Duschka, F. M. Vogt, J. Barkhausen, and T. M. Buzug. A hybrid dual-axis tilt acquisition geometry for digital musculoskeletal tomosynthesis. *Phys. Med. Biol.*, submitted:1–14, 2013. 137

[11] Y. M. Levakhina, J. Mueller, R. L. Duschka, F. M. Vogt, J. Barkhausen, and T. M. Buzug. Weighted simultaneous algebraic reconstruction technique for tomosynthesis imaging of objects with high-attenuation features. *Med. Phys.*, 40:1–12, 2013. doi: 10.1118/1.4789592. 100, 120

Co-author

[12] A. Cordes, Y. M. Levakhina, and T. M. Buzug. Mikro-CT basierte Validierung digitaler Tomosynthese Rekonstruktion. In *Bildverarbeitung für die Medizin*, pages 304–309, 2012. 17, 161

[13] A. Cordes, Y. M. Levakhina, and T. M. Buzug. A method for validation and evaluation of digital tomosynthesis reconstruction. In *Biomed Tech*, page 513, 2012. 17, 161

[14] S. Kiencke, Y. M. Levakhina, and T. M. Buzug. Greedy projection access order for SART (Simultaneous Algebraic Reconstruction Technique). In *Bildverarbeitung für die Medizin*, pages 93–98, 2013. 95

[15] J. Mueller, F. Kaiser, Y. M. Levakhina, M. Stille, I. Weyers, and T. M. Buzug. An open database of metal artifacts cases for clinical CT imaging. In *Proceedings of The Second International Conference on Image Formation in X-Ray Computed Tomography*, pages 210–213, 2012.

[16] R. L. Duschka, Y. M. Levakhina, L. C. Busch, P. Hunold, T. M. Buzug, J. Barkhausen, and Vogt F. M. Tomosynthese zur Beurteilung des Handskeletts - technische Möglichkeiten und klinisches Potenzial. In *Fortschr Rontgenstr*, volume 182, page A25, 2010.

[17] R. L. Duschka, P. Bischoff, Y. M. Levakhina, P. Hunold, T. M. Buzug, L. C. Busch, J. Barkhausen, and F. M. Vogt. Tomosynthesis - the revival of an imaging tool for musculoskeletal applications. In *ECR 2011 - European Congress of Radiology, Wien, Austria*, pages C–1895, 2011.

[18] R. L. Duschka, Y. M. Levakhina, T. M. Buzug, P. Hunold, J. Barkhausen, and F. M. Vogt. Tomosynthesis: A long forgotten imaging tool becomes a fierce competitor to cr in follow-up examinations of pathologies in hands. In *RSNA - 97th Scientific Assembly and Annual Meeting, Chicago*, pages LL–MKS–TU2A, 2011.

1.4 Publications

[19] R. L. Duschka, P. Bischoff, K. May, Y. M Levakhina, T. M. Buzug, A. Kovacs, P. Hunold, J. Barkhausen, and F. M. Vogt. Digitale Tomosynthese - Ein neues Verfahren zur Beurteilung degenerativer Gelenkveränderungen im Vergleich zum konventionellen Röntgen. In *RöFo: Fortschritte auf dem Gebiet der Röntgenstrahlen und bildgebenden Verfahren*, volume 184-VO216 4, 2012.

[20] R. L. Duschka, P. Bischoff, K. May, Y. M. Levakhina, T. M. Buzug, A. Kovacs, P. Hunold, J. Barkhausen, and F. M. Vogt. Digital tomosynthesis: a fierce competitor to cr in examinations of small bones. In *ECR 2012 - European Congress of Radiology, Wien, Austria*, pages C–1514, 2012.

Chapter 2

Where we are today: Tomosynthesis research and development

Contents

2.1	Tomosynthesis basics	12
2.2	From radiostereoscopy to digital tomosynthesis	18
2.3	Evolution of reconstruction algorithms	23
2.4	CT and tomosynthesis today: practical comparison	26
2.5	Problems of limited data tomography	33

For many centuries physicians wished for a method to look inside patient without the need for a surgical intervention. The level of recognition of human diseases would have greatly increased if there was a way to make a human body "transparent". Nobody could have even imagined that this dream can come true. With the discovery of X-rays by Wilhelm Conrad Röntgen in 1895 (Roentgen 1895a, Roentgen 1895b) it became possible.

Nowadays, X-ray based imaging techniques include a variety of implementations and applications. Computed radiography (CR)and digital radiography (DR)are used for planar imaging when a three-dimensional object is mapped onto a two-dimensional plane. Variations of computed tomography (CT), such as clinical CT, C-arm, tomosynthesis, micro-CT and industrial CT are used to "cut" an object into a stack of tomographic slices. These X-ray based imaging methods are widely used not only for diagnostics and assistance in clinical practice but also for screening in security applications and for non-destructive material testing in industry, archeology and material sciences.

The aim of this chapter is to give an introduction into a modern X-ray based imaging technique, called digital tomosynthesis (DT). First, the basic principles will be explained. Second, a historical overview of the imaging principle and the reconstruction algorithms will be given. Afterwards, a comparison of state-of-the-art devices will be presented. Finally, the theoretical aspects of DT will be discussed and open questions and problems will be summarized.

2.1 Tomosynthesis basics

The basics of the tomosynthesis imaging technique will be given in this section.

2.1.1 Introduction

DT is an X-ray based tomographic imaging technique. It is a non-invasive and non-destructive method for the three-dimensional visualization of the inner structures of an object. DT is known as an attractive low-dose alternative to CT in medical and non-medical imaging applications, when the data acquisition over the full angular range is impossible or infeasible if object is too large or if only a small part of the object is of interest. The primary application of DT is the screening for breast cancer. Here it is used together with traditional mammography for the detection of microcalcifications and tumors (Niklason 1997, Park 2007, Baker 2011). Further medical application fields include pulmonary nodules detection in chest imaging (e.g. Dobbins 2008, Tingberg 2010), dental imaging (Ogawa 2010) and musculoskeletal imaging of hands (Duryea 2003). Non-medical applications of tomosynthesis include security luggage screening in airports (Reid 2011) and non-destructive material testing in industrial imaging (Huang 2004).

A DT data acquisition includes measuring a limited number of low-dose two-dimensional projections of an object. This is done by moving a detector and/or an X-ray tube around the object within a limited angular range. Each measured two-dimensional *intensity* image $I(\mathbf{u}, \theta)$ represents the decreased signal. The decrease is caused by photon-matter interactions (photoelectric absorption, scatter). If no object is present, the initial intensity I_0 will be measured. The model of tomosynthesis measurements is based on the Beer-Lambert absorption law for a polyenergetic spectrum of the X-ray tube and additionally includes a scatter term $r(\mathbf{u}, \theta)$ and the detector efficiency $\varepsilon(E)$

$$I(\mathbf{u}, \theta) = \int_{E_{max}} \varepsilon(E) I_0(E) e^{-\int_L \mu(\mathbf{x}, E) d\mathbf{x}} dE + r(\mathbf{u}, \theta). \qquad (2.1)$$

Here, L is a path through the object, depending on the direction of the beam θ, $\mathbf{u}$ is a vector describing a point on the detector and $\mu(\mathbf{x}, E)$ is the distribution of X-ray

2.1 Tomosynthesis basics

attenuation coefficients (measured in cm^{-1}) in dependency of position **x** within the imaged volume and the energy E (measured in kilo electron volts keV). For simplicity, the scatter term and the detector efficiency term are usually not taken into account. As a further simplification, it can be assumed that the X-ray spectrum is a monoenergetic spectrum. Thus, the dependency of μ on the energy can be omitted. The resulting model is the well known exponential Beer-Lambert law for a monoenergetic spectrum

$$I(\mathbf{u}, \theta) = I_0 e^{-\int_L \mu(\mathbf{x}) d\mathbf{x}}. \tag{2.2}$$

A *projection* image $p(\mathbf{u}, \theta)$ has a linear relation to the attenuation coefficients $\mu(\mathbf{x})$ and is defined as the log-transform

$$p(\mathbf{u}, \theta) = -log\left(\frac{I(\mathbf{u}, \theta)}{I_0}\right) = \int_L \mu(\mathbf{x}) d\mathbf{x}. \tag{2.3}$$

Two exemplary images of a hand measured using a medical tomosynthesis device are shown in Fig. 2.1a (intensity image) and Fig. 2.1b (projection image). Tomosynthesis projection images can also be post-processed for a better visual perception (see Fig. 2.1c).

The task of DT is to reconstruct an unknown distribution of X-ray attenuation coefficients $\mu(\mathbf{x})$ within the imaged object based on the limited set of measured line integrals $p(\mathbf{u}, \theta)$, i.e. to solve an ill-posed inverse problem. Depending on the type of reconstruction algorithm, either intensity images or projection images are used for reconstruction. In DT the reconstructed slices are typically parallel to the detector, see Fig 2.2a and Fig. 2.2b. Reconstructed tomosynthesis slices of an apple and of a hand at different heights are shown in Fig. 2.2c-Fig. 2.2h. Three reconstructed slices of an apple at 20 mm, 30 mm and 40 mm heights show different cuts through the seeds of the apple. The reconstructed slices of the hand show that different regions of bone are sharp at different heights (marked by ellipses). Regions with distal phalanges are shown at 10 mm. A region with proximal phalanges and a region with carpal bones are shown at 20 mm and a region with metacarpal bones is shown at 28 mm.

Reconstructed images show that despite the fact that a tomosynthesis dataset consists only of a limited number of projections acquired over a limited angular range, the reconstruction of structures at their correct geometrical location is possible.

Chapter 2. Tomosynthesis today

(a) intensity image

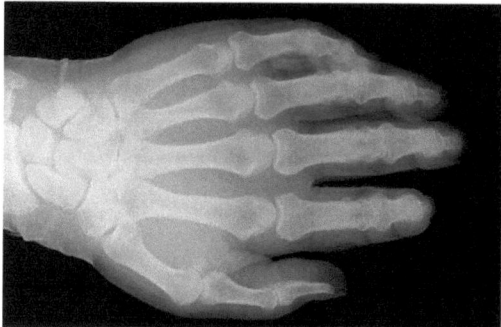

(b) projection image

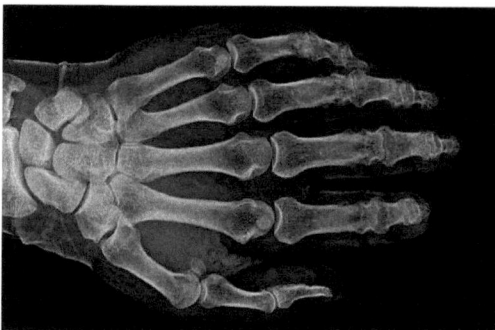

(c) processed projection image for better visual perception

Figure 2.1: Tomosynthesis raw-data of a hand acquired using Siemens Mammomat Inspiration device. (a) intensity image; (b) projection image; (c) post-processed projection image for better visual perception.

2.1 Tomosynthesis basics

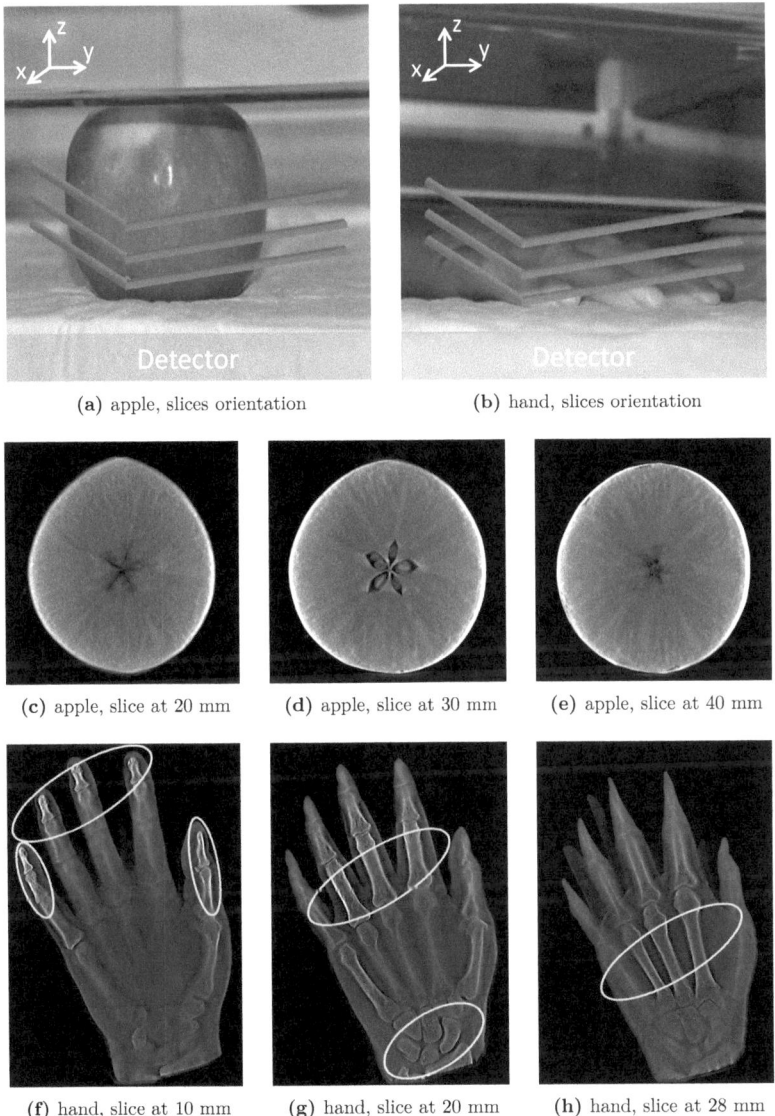

Figure 2.2: Orientation of slices in tomosynthesis (a-b) and tomosynthesis reconstruction results of an apple (c-e) and a hand (f-h). The total height of the apple is 60 mm (60 slices). The total height of the hand is 35 mm (35 slices). Slices at different heights show different (anatomical) structures.

2.1.2 Tomosynthesis technology

2.1.2.1 Acquisition geometry

A tomosynthesis device typically consists of an X-ray tube and a flat-panel detector. During a tomosynthesis acquisition they are moved along a pre-defined trajectory. Several types of motion are possible: a parallel path, a full and a partial isocentric and a circular geometry. In the linear path geometry (Fig. 2.3a) the X-ray tube and the detector are synchronously moved along a line in opposite directions. This geometry corresponds historically to the analogue conventional geometric tomography, see e.g. (Ziedses des Plantes 1932). In the full isocentric geometry (Fig. 2.3b) the X-ray tube and the detector are moved along an arc trajectory around a common origin. If the tube and the detector are moved over 360°, this geometry describes a cone-beam tomographic system. In the partial isocentric geometry the X-ray tube is moved along an arc while the detector is moved along a line (Fig. 2.3c) or stays fixed (Fig. 2.3d). The partial isocentric geometry with the fixed detector is typically used for breast imaging. In the circular geometry (Fig. 2.3e) the tube and the detector are moved in parallel planes along a circular trajectory. Such a geometry is also known from early works on tomosynthesis (Grant 1972) and is nowadays used in industrial applications. More information about geometries of motion in tomosynthesis can be found e.g. in the review paper by J. T. Dobbins (Dobbins 2003).

2.1.2.2 Geometry parameters

The tomosynthesis geometry parameters include the angular range, the number of projections and the angular step size (a projection density). Parameters for the geometry with a fixed detector and an X-ray tube moving along an arc trajectory are shown schematically in Fig. 2.3f. The *angular range* of the X-ray tube rotation or the *sweep angle* is denoted by θ. In case of a partial isocenter geometry it is defined as the size of the total arc around the rotation center and is described by the X-ray tube position from the first measured projection to the last measured projection. In case of a circular trajectory it is defined as a two-dimensional angle in a three-dimensional space (solid angle) drawn by the X-ray tube. The angular range in clinical applications is typically between 20° and 50°. The *number of projections*, denoted by N_{proj}, is the number of measured X-ray images acquired over the angular range θ. The number of projections in clinical applications is typically between 10 and 30. The *angular step size*, denoted by $\Delta\theta$, is defined as the total angular range divided by the number of projections and is described as the angle between the current and the next position of the X-ray tube. Sometimes an inverse measure, called the *projection density*, can be found in

2.1 Tomosynthesis basics

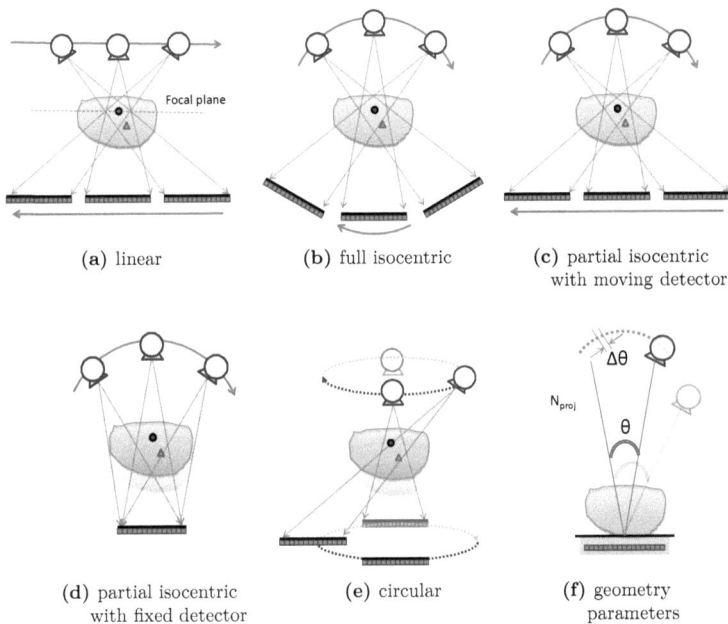

(a) linear (b) full isocentric (c) partial isocentric with moving detector

(d) partial isocentric with fixed detector (e) circular (f) geometry parameters

Figure 2.3: (a)-(e) Tomosynthesis geometries; (f) a schematic illustration of parameters for a device with a fixed detector and an X-ray tube moving along an arc trajectory. The following parameters are shown: the angular range θ, number of projections N_{proj} and the angular step $\Delta\theta$.

the literature (Deller 2007, Machida 2010). The projection density is the number of projections divided by the angular range. Another parameter, which is not directly related to the technical parameters of the device but can be seen as a geometry parameter, is the object orientation in the detector xy-plane. It plays an important role and influences the image quality (Cordes 2011, Cordes 2012a, Cordes 2012b, Cordes 2013).

2.1.3 DICOM format

The Digital Imaging and Communications in Medicine (DICOM) standard is a global information-technology standard created by the National Electrical Manufacturers Association (NEMA). DICOM is used for distributing and viewing any kind of medical images worldwide (webNEMA 2013). A DICOM file contains the image data and additionally a file header with supplementary information on the patient, the study, the device, the physician and the image. There is a number of free DICOM viewers available to open DICOM files, visualize two-dimensional and three-dimensional data and read the header information, among them are ImageJ, Sante DICOM viewer and Agnosco DICOM Viewer. The interested reader can refer, e.g. to the web-page of Chris Rorden (Rorden, C. 2013) for a list of links to free DICOM viewers. There are three build-in MATLAB® commands to read and to save DICOM data and DICOM header automatically: `dicomread`, `dicominfo` and `dicomwrite`.

Typically, the tomosynthesis raw-data and reconstructed images can be exported as a set of DICOM files. If tomosynthesis raw-data are available as DICOM, all necessary information for simulation and reconstruction can be found in the header, see Table 2.1. A DICOM header is comprised of DICOM elements. A DICOM element has a tag, a data type, a length and a value. The tag uniquely defines the properties of an element. It consists of two groups of four digits (sometimes letters), separated by comma and called a Group and an Element. Each attribute has also a name. Typically attribute names are self explaining. A lookup table with the explanation of DICOM attribute names and the corresponding tag numbers can be found, e.g. at (webDICOMLOOKUP 2013). According to the table, e.g. the tag (0018,1110) has the attribute name Distance Source to Detector and denotes the distance in mm from the source to the detector center.

Table 2.1: Important tomosynthesis-related DICOM tags.

Tag	Attribute name
(0018,0050)	Slice Thickness
(0018,11A0)	Body Part Thickness (object height)
(0018,1110)	Distance Source to Detector
(0018,1111)	Distance Source to Patient
(0018,1530)	Detector Primary Angle (X-ray tube angle)
(0018,7026)	Detector Active Dimension(s)
(0028, 0010)	Rows (detector)
(0028, 0011)	Columns (detector)

2.2 From radiostereoscopy to digital tomosynthesis

The principle of *stereoscopy* has been proposed even before the discovery of X-rays. In 1838 Sir Charles Wheatstone demonstrated the physics of binocular vision and proposed a novel device which he called a stereoscope (Wheatstone 1838). The device was designed to present two images with a slight angular offset separately to the left

2.2 From radiostereoscopy to digital tomosynthesis

and the right eye of an observer and by this to create a three-dimensional impression of the presented scene. Moreover, Sir Wheatstone also mentioned in his paper that the discussion on the different visual impression of an object itself and a painting of this object can be already found in the Trattato della Pittura (Treasure of Paining, 1721) of Leonardo da Vinci. The stereoscopy principle was used in the very early development of radiology applications (e.g. Thomson 1896). Apparently, one can draw a connection between the stereoscopy and modern tomosynthesis imaging principle.

The actual history of X-ray imaging starts in 1985 when Wilhelm Conrad Röntgen discovered a new kind of radiation which he called "X-rays" (Roentgen 1895a, Roentgen 1895b). It was a breakthrough invention, which made possible the visualization of inner details of a human body without a surgical intervention. Already shortly after the discovery of X-rays, the harmful effect of ionizing radiation was observed, which cause radiation injury (Upton 1992). Despite all the risks, the X-ray imaging still offers very attractive opportunities.

A simple radiograph contains the superposition of all three-dimensional structures in an object as a two-dimensional shadow. As a consequence, it is impossible to recover the information from which exact three-dimensional position any particular feature (e.g. tumor) originates. In the beginning of the 1920s there were many attempts to erase superimposed shadows from X-ray images and to benefit from the use of X-rays for imaging of the human body. This resulted in a number of patent applications, e.g. in 1922 from the French scientist A. E. M. Bocage (Bocage 1822) and in 1927 from the German scientist E. Pohl (Pohl 1927). A work from the Franco-American technologist J. Kieffer (Kieffer 1929) was also patented and later was commercialized. Besides patents, there was also a number of papers published, among them e.g. in 1914 by the Polish scientist K. Mayer (Mayer 1916), in 1930 by the Italian scientist A. Vallebona (Vallebona 1932) and in 1932 by the Dutch engineer B. G. Ziedses des Plantes (Ziedses des Plantes 1932). Owing to the fact that the communication between researchers from different countries was very limited at that time, all those scientists rediscovered similar concepts. As a summary, all those works deal with the same imaging technique, in which the X-ray tube and the detector are moved in two parallel planes. The goal was to display the plane in focus very sharp and to blur the planes that are out of focus. This technique became known under several names: it was called a *stratigrafia* by A. Vallebona, a *laminography* by J. Kieffer and a *planigraphy* by A. E. M. Bocage and B. G. Ziedses des Plantes. In 1935 G. Grossmann presented a device which he called a *tomograph* (Grossmann 1935a, Grossmann 1935b). It should be noted that the Grossmann's tomograph had nothing to do with the modern tomographic devices but it was also based on the abovementioned principle. More information can be found e.g. in the historical article written by the curator of the Belgian Museum of Radiology R. van Tiggelen (Van Tiggelen 2002).

The next evolutionary step was the implementation of a device, that enabled the storage of each measured radiograph as a set of separate analogue images. The stored images were processed after the examination instead of doing an integration of measurements directly on the film. Using a set of measured radiographs, it is possible to generate an arbitrary number of planes (Garrison 1969) or laminograms (Miller 1971) through the object. The total radiation dose can thus be reduced because

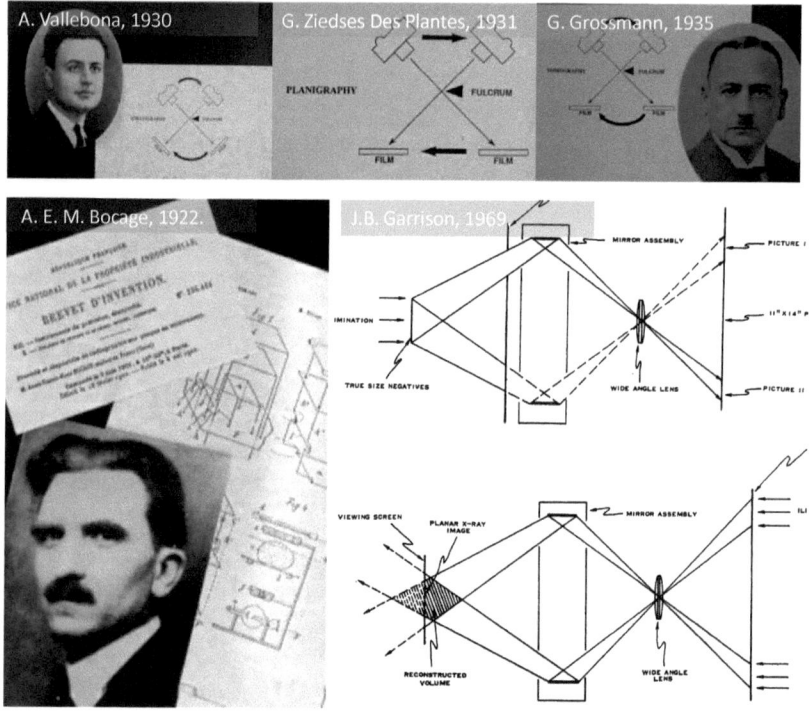

Figure 2.4: A photo collage showing photos of A. E. M. Bocade, G. Ziedses des Plantes, A. Vallebona, G. Grossmann and sketches of early works on tomosynthesis. (Photos courtesy of R. van Tiggelen. Schematical illustrations from (Garrison 1969) are reprinted with permission from the American Journal of Roentgenography.)

2.2 From radiostereoscopy to digital tomosynthesis

only one examination is needed to produce images of the whole volume. This is essentially the main idea of the modern tomosynthesis as it is know nowadays. The word *tomosynthesis* itself has been introduced a year later in 1972 by D. G. Grant (Grant 1972). Figure 2.4 shows some sketches from the first papers on conventional tomography and tomosynthesis and a collage with photos of A. E. M. Bocade, G. Ziedses des Plantes, A. Vallebona, G. Grossmann. A number of further improvements of tomosynthesis has been proposed during the 1970s and 1980s. It resulted in different variations, including, e.g. ectomography and flashing tomosynthesis. All changes concerned mainly the improvement of the image quality to suppress out-of-plane details and the shortening the acquisition time. Review papers by Dobbins give a detailed overview over the tomosynthesis research at that time (Dobbins 2003, Dobbins 2009).

In the same year (1972) another significant event took place. Sir Godfrey Hounsfield and James Ambrose gave a talk on "Computerised Axial Tomography" at the 32nd Congress of the British Institute of Radiology. They presented first tomographic clinical scans acquired using the head scanner named after the company Electric and Musical Instruments (EMI). The images were produced using the "true" tomographic principle with a full rotation around the patient as it is known nowadays. For more information about the EMI head scanner see the original papers from G. Hounsfield (Hounsfield 1973) and J. Ambrose (Ambrose 1973) and e.g. the following review papers (Beckmann 2006, Gould 2009). This development resulted in a decrease of the interest in the tomosynthesis technology due to the obvious advantages of CT over conventional tomography and tomosynthesis. According to J. T. Dobbins, tomosynthesis was forgotten in the late 1980s with the development of spiral CT. Table 2.2 summarizes the milestones in development of analogue tomosynthesis from radio-stereometery to the spiral CT.

Tomosynthesis has gained a renewed interest in the beginning of the 21st century due to advances in digital flat-panel detector technologies (Dobbins 2009). The combination of fast and large digital flat-panel detectors and improved computer technologies offered an attractive solution to the problem of long examination and processing time in tomosynthesis and brought this technique into focus of research again. The number of papers on tomosynthesis indexed in PubMed (webPUBMED 2013) per year is growing almost exponentially (see Fig. 2.5). In February 2011 the first tomosynthesis device Selenia Dimensions 3D System by Hologic, Inc. has been approved by the U. S. Food and Drug Administration (FDA) (webFDA 2013). Siemens and other manufacturers are currently waiting for approval.

Table 2.2: Out of the shadows - first paper on tomosynthesis.

Year	Name	Technique
1915	C. Baese	radio-stereometer
1921	P. A. Bocage	radiographic stereoscopy
1921	B. G. Ziedses Des Plantes	planigraphy
1928	J. Kieffer	laminagraphy
1930	A. Vallebona	stratigraphy
1934	H. Chaoul, G. Grossmann	tomography
1947	R. Sans and J. Porcher	polytome
1969	J. B. Garrison, E. R. Miller	infinite number of laminograms
1972	D. G. Grant	tomosynthesis
1972	G. Hounsfield and J. Ambrose	Computerised Axial Tomography (EMI scanner)
1989	W. Kalender	Spiral CT

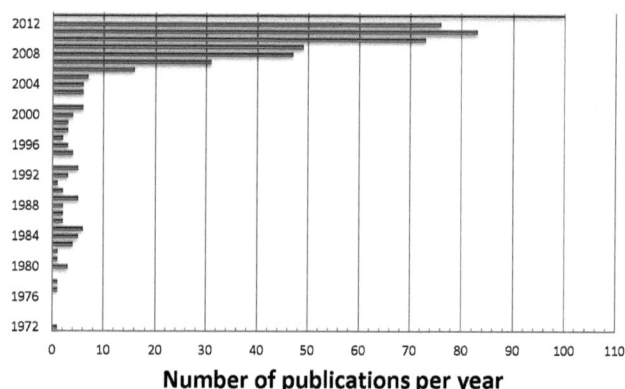

Figure 2.5: Trend in tomosynthesis research: Number of scientific papers on tomosynthesis indexed in PubMed (webPUBMED 2013) per year. The MATLAB® code for PubMed trend search is presented in Appendix.

2.3 Evolution of reconstruction algorithms

In contrast to CT, an accurate image reconstruction in DT is a challenging task because only a limited number of projections acquired over a limited angular range is available. Thus, limited angle artifacts are unavoidable. They appear as blurred copies of objects located in other slices. The artifacts decrease the diagnostic relevance of the DT images. Various reconstruction approaches to reduce artifacts in tomosynthesis images have been proposed so far.

The historically first tomosynthesis system did not require any reconstruction algorithms because the image was formed directly on film cassettes during the measurement process. In 1965 J. B. Garrison, E. R. Miller and D. G. Grant proposed to store the projection images separately and post-process them using a backprojection-like algorithm, named playback geometry, see e.g. (Garrison 1969). Regardless of the acquisition type, it was possible to focus on a certain plane and obtain it sharp. The focusing was done mechanically by positioning the X-ray tube and the image receptor before the measurement process begins or afterwards using a semitransparent mirror in a reprojection system. Objects, which belong to this plane are always projected onto the same location and therefore coincide and appear sharp, while details from the out-of-focus planes are superimposed and blurred. This method was based on a simple geometrical principle and had an inherent backprojection blur. The geometry of motion defined the degree and the direction of the blur. The main role of the blur was the effacement of unwanted details outside the focal plane while preserving the details of the in-focus plane (Littleton 1965). In fact, blurring cannot accomplish this task and is always seen by the human eye as a disturbing component. Therefore, a variety of tomosynthesis deblurring methods has been proposed.

The most simple deblurring strategy was developed by P. R. Edholm in 1969 for analogue images (Edholm 1969). He proposed to subtract the blur using a blurred transparent negative photographic copy of the original image. His formula is easily understandable

$$\begin{aligned}\text{(measured image)} &+ \text{(blurred negative copy)} &=& \text{(improved image);} \\ \text{(sharp details + blur)} &+ \text{(- blur)} &=& \text{(sharp details).}\end{aligned} \quad (2.4)$$

Later, the same method was applied by D. P. Chakraborty to digital tomosynthesis and was called self-masking subtraction tomosynthesis (Chakraborty 1984). It was, in general, identical to the one-dimensional high-pass spatial frequency filtering. The suppression of the low frequencies reduces not only the blur but also the low frequencies from the object itself. Therefore, there was an attempt to use a band-pass filter (Sone 1991a) or to apply an additional two-dimensional unsharp mask (Sone 1991b).

U. E. Ruttiman (Ruttimann 1984, van der Stelt 1986) proposed a novel concept to use planes adjacent to the plane of interest and an iterative spatial deconvolution. Since blur is produced by the features in out-of-focus planes, a set of properly convolved (blurred) adjacent planes can be subtracted from the plane of interest to remove an undesired contribution of details lying outside the target plane. A similar approach, called selective plane removal, was proposed by Gosh Roy in 1985 (Roy 1985). The blurring function (point spread function) must be considered to develop the relevant equation for eliminating distortions from the closely adjacent planes. Another approach, which was also called selective plane removal (Kolitsi 1993), uses a preliminary reconstructed volume (a set of planes) to reproduce the blur on the target plane, arising from the other reconstructed planes. Each plane is reprojected onto the target plane for each view in order to synthesize the blur from the unrelated structures and eliminate it. An extended approach, called matrix inversion tomosynthesis (MITS) (Godfrey 2006), uses linear algebra to solve the blurring problem in each plane.

A non-linear reconstruction method, called extreme-value decoding, was proposed for task-specific applications, for example in angiography, when only the projection with the largest or smallest value is used (Haaker 1985a, Haaker 1985b, Stiel 1993). A voting strategy, which rejects the projections that include too high values, was used together with Maximum Likelihood (ML) algorithms (Wu 2006) for the reduction of artifacts caused by a metal needle in breast tomosynthesis. This method works only if high-attenuation features can be detected and accurately segmented. Other methods to reduce blurring use wavelets (Badea 1998) or three-dimensional anisotropic diffusion filtering (Sun 2007). An empirical adaptive weighting scheme which allows for the simple backprojection blur reduction in musculoskeletal tomosynthesis has been presented recently (Levakhina 2012a). As a conclusion, simple backprojection algorithms are computationally inexpensive and provide images with good noise properties but poor contrast. They are typically combined with different post-filtering and deblurring algorithms. Alternatively, non-linear backprojections have also been proposed.

Filtering is applied not only in the image domain but also in the projection domain before the summation is done. This is actually equivalent to the well-known filtered backprojection (FBP) algorithm. For example, the *ectomography* method, presented by P. R. Edholm, C. U. Petersson and H. E. Knutsson in 1980 (Edholm 1980, Petersson 1980, Knutsson 1980) applies a high-pass and a low-pass filter to the projection images in perpendicular directions before summation. Nowadays, the FBP is widely used

2.3 Evolution of reconstruction algorithms

reconstruction method in clinical routine in full angle CT. This algorithm has the advantage of the high computational performance. In the limited angle case, however, it suffers from the missing regions in the Fourier space. A number of papers are exploring the idea of improving the sampling in Fourier space or designing a proper filter for the tomosynthesis projections (Matsuo 1993, Lauritsch 1998, Stevens 2001, Claus 2006, Ludwig 2008).

Iterative reconstruction algorithms have also been adopted from CT. One major benefit of iterative methods over FBP is the possibility to include prior knowledge (a positivity constraint, an object extend, a noise model) into the reconstruction process. Iterative methods can be easily adapted to a new imaging geometry, e.g. non-circular X-ray tube trajectories, non-uniform angular spacing or a fixed detector. While only forward- and backprojectors must be adapted for iterative reconstruction, a new appropriate filter must be derived analytically for FBP for each geometry. Both, algebraic and statistical iterative reconstruction algorithms provide acceptable image quality in breast tomosynthesis when the reconstruction parameters are properly chosen (Zhang 2006b). In (Levakhina 2011a) and (Levakhina 2011b) it was shown that algebraic reconstruction is capable of suppressing some structural noise caused by bone tissue and results in a improved reconstruction for hand imaging. Despite these advantages, the problem of data insufficiency still remains for iterative reconstruction algorithms. A detailed review of reconstruction methods as well as pre- and post-processing algorithms for tomosynthesis can be found in literature (Dobbins 2003). A detailed comparison of reconstruction algorithms can be found e.g. in (Wu 2004) and (Zhang 2006b).

To sum up, a list of reconstruction algorithms for tomosynthesis is given below:

1. Deblurring algorithms
   - 1D high-pass or band-pass spatial frequency filtering in image space;
   - reproducing blur from adjacent planes (selective plane removal, constrained iteration method);
   - Matrix Inversion Tomosynthesis (MITS);
   - task-dependent: extreme value decoding, voting strategy;
   - wavelets-based deblurring;
   - three-dimensional anisotropic diffusion filtering.

2. FBP
   - ectomography;
   - filter adaptation;
   - completion of Fourier space;

3. Iterative algorithms.

2.4 CT and tomosynthesis today: practical comparison

As has been mentioned in previous sections, DT and CT are closely related. They share the common physical principle and reconstruction algorithms. However, there are a number of major differences when it comes to the comparison of the actual devices. The aim of this section is to compare a state-of-the-art breast tomosynthesis device with a clinical CT device and a micro-CT device based on the example of the Siemens Mammomat Inspiration, Siemens Somatom Definition AC and the Skyscan 1172 micro-CT. Photos of those devices are shown in Fig. 2.6. They will be compared in terms of technical parameters and reconstructed images. Some images acquired with a CR device will be shown as well for an additional comparison.

2.4.1 Technical parameters

The tomosynthesis Siemens Mammomat Inspiration device is equipped with a half-cone X-ray tube, a fixed detector, an object-support table and a compression paddle. The tube path, the iso-center and the chest-side of the detector are located in one plane, which is oriented perpendicular to the detector plane. The rotation iso-center is located close to the detector, which results in a magnification factor slightly larger than one. This is used to avoid truncation artifacts caused by a large magnification factor when the object is projected outside the detector sensitive area. The compression paddle is needed to compress the examined breast and to avoid motion blur. During an acquisition the X-ray tube moves along a 50°-arc and the stationary detector acquires 25 low-dose projection images with 2° angular step size. The whole acquisition takes approximately 30 seconds. The side and the front views of the imaging geometry are shown in Fig. 2.7. A three-dimensional schematic illustration of the Siemens Mammomat Inspiration geometry is shown in Fig. 2.8.

In CT, in contrast to DT, a full angular dataset is acquired. This is done by rotating either the gantry in a clinical CT or the object in a micro-CT. The typical number of acquired projections is more than a thousand and the angular step size is smaller than 0.3^o. The acquisition time varies from less than one second in CT to up to several hours in micro-CT. The total radiation dose is higher than in the DT case. A curved multi-row detector is used in CT, while a flat-panel detector is used in both, micro-CT and DT devices. The resulting DT raw dataset is about 0.5 GB and it can be up to several GB in CT and micro-CT. A list of technical parameters is given in Table 2.3.

2.4 CT and tomosynthesis today: practical comparison

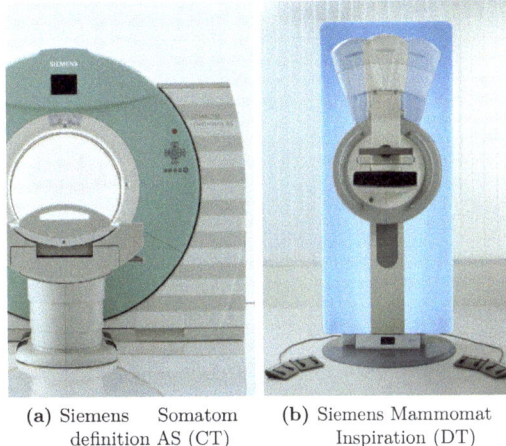

(a) Siemens Somatom definition AS (CT)

(b) Siemens Mammomat Inspiration (DT)

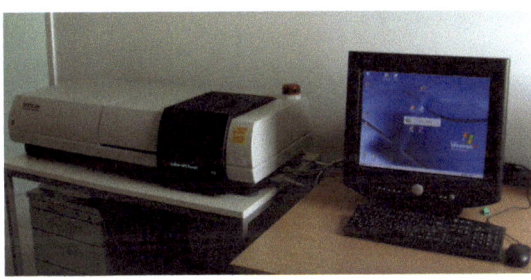

(c) Skyscan 1172 (micro-CT)

Figure 2.6: Photos of the Siemens Somatom definition AS CT device (a), the Siemens Mammomat Inspiration tomosynthesis device (b) and the Skyscan 1172 micro-CT device (c). ((a) and (b) Source: Siemens Healthcare).

2.4.2 Reconstructed images

In full angle tomographic applications, slices are reconstructed perpendicular to the detector plane, while in DT slices are typically reconstructed parallel to the detector plane. The slice orientation direction in DT is due to the fact that the limited angular range limits the in-depth resolution of the reconstructed images. The clinical CT provides an isotropic resolution ($0.6 \times 0.6 \times 0.6$ mm), which is not high enough to resolve microcalcifications in breast imaging or fine trabecular bone structures in the imaging of

28   Chapter 2. Tomosynthesis today

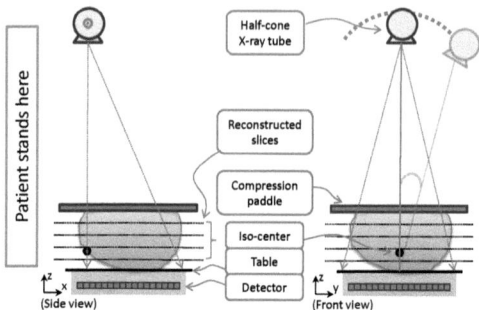

Figure 2.7: A schematic illustration of the tomosynthesis Siemens Mammomat Inspiration imaging geometry. The detector is fixed and the X-ray tube moves along an arc trajectory. The orientation of the reconstructed slices is parallel to the detector. Front and side views of the device are shown.

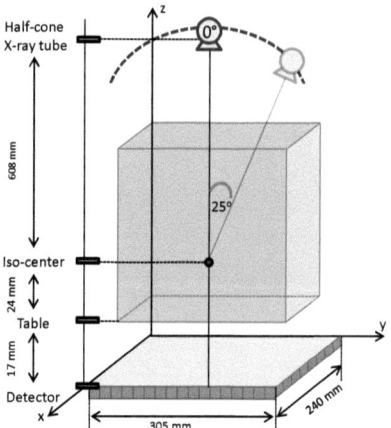

Figure 2.8: Three-dimensional schematic drawing of the Siemens Mammomat Inspiration geometry.

hands. The micro-CT also provides an isotropic resolution up to $20 \times 20 \times 20$ µm, which is better than the resolution of CT. Hoverer, the Skyscan 1172 micro-CT is not suitable for in-vivo imaging because of the long acquisition time and the limited maximum object size $(4\ \mathrm{cm}^3)^1$. DT provides anisotropic resolution, i.e. the in-plane resolution is high $(0.085 \times 0.085$ mm), while the axial resolution is limited (1 mm). Besides tomographic

---
[1] micro-CT devices exist, which are suitable for in-vivo animal imaging, e.g. SkyScan 1076

2.4 CT and tomosynthesis today: practical comparison

techniques, a two-dimensional CR can be used for structure visualization inside an object. It has the same high in-plane resolution as tomosynthesis but has a poor contrast and provides no three-dimensional information because of the averaging process. A schematic illustration of the resolution comparison of CR, DT and CT is shown in Fig. 2.9.

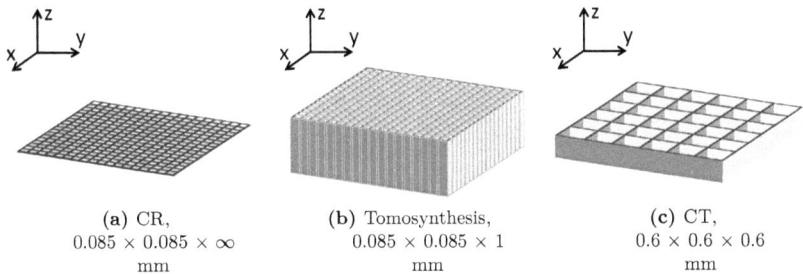

(a) CR,
0.085 × 0.085 × ∞
mm

(b) Tomosynthesis,
0.085 × 0.085 × 1
mm

(c) CT,
0.6 × 0.6 × 0.6
mm

Figure 2.9: Illustration showing a resolution comparison of CR (a), DT (b) and CT (c). CR and DT have a high in-plane resolution. CR provides no three-dimensional information and DT has a non-isotropic resolution with a limited z-resolution. CT has isotropic resolution.

The images of a hand produced using CR, CT and DT are shown in Fig. 2.10. In Fig. 2.10b-Fig. 2.10d a finger region of interest (ROI) is shown and in Fig. 2.10f-Fig. 2.10h a carpal bone (wrist) ROI is shown. The images confirm visually the abovemetioned effect of the resolution and contrast. The resolution of the CT images (Fig. 2.10d and Fig. 2.10h) is not high enough to clearly visualize fine trabecular structures and bone margins. Images produced using CR (Fig. 2.10b and Fig. 2.10f) and DT (Fig. 2.10c and Fig. 2.10g) have the same resolution, but CR images have a worse contrast. Additionally, some limited angle artifacts are present on DT images, which will be discussed later.

Another comparison of reconstructed images is shown in Fig. 2.11. Here, three micro-CT slices (Fig. 2.11b-Fig. 2.11d) and DT slices (Fig. 2.11e-Fig. 2.11g) of a dried finger bone are shown. The presented slices are located at approximately the same height. The micro-CT slices have been averaged to the slice thickness of 1 mm in order to make them visually comparable to the DT slices. Despite the limited in-depth resolution of DT, structures are reconstructed in their correct locations. Almost all structures, which are presented in micro-CT images can be recognized in DT images. A comparison of the image reconstruction parameters of CT, micro-CT and DT is presented in Table 2.4.

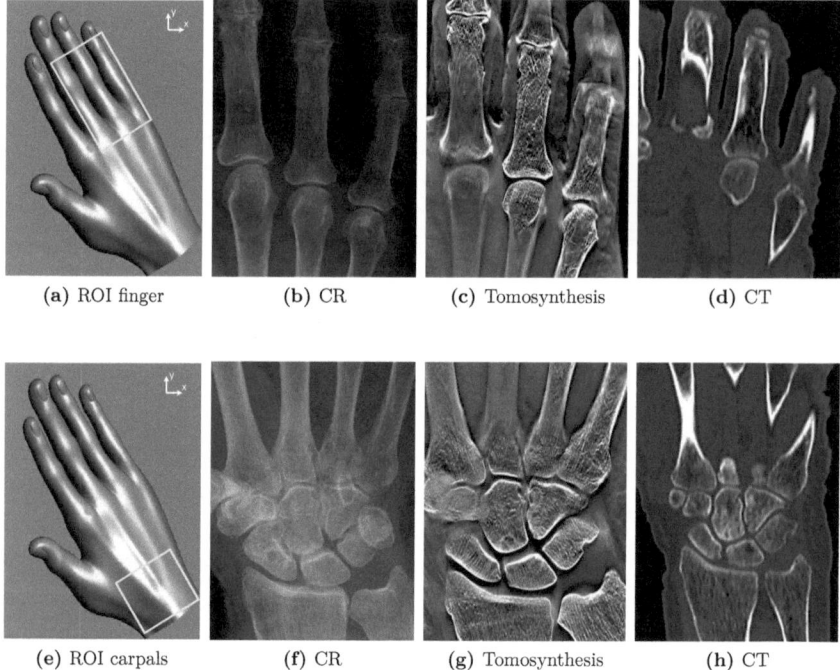

Figure 2.10: Comparison of CR, DT and CT images for region of interest containing fingers (b)-(d) and carpal bones (wrist) (f-h).

2.4 CT and tomosynthesis today: practical comparison

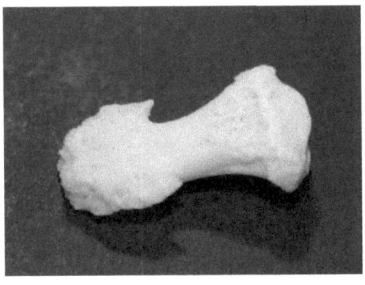

(a) photo of a finger bone

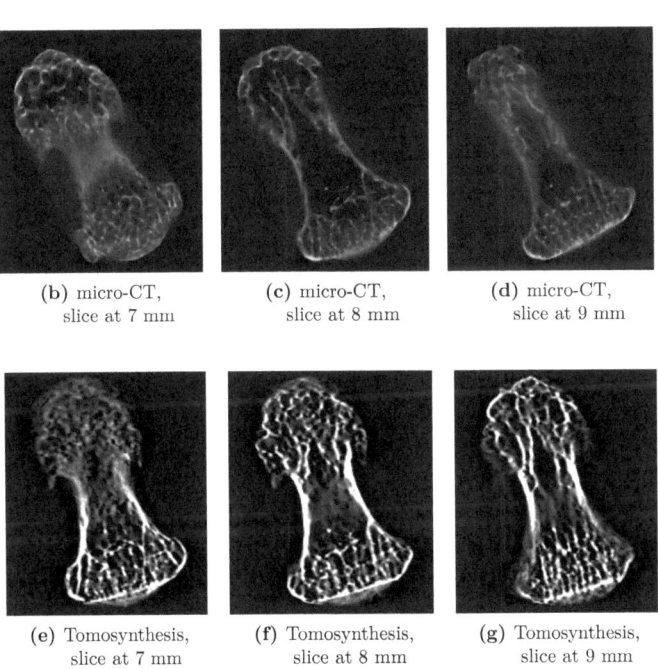

(b) micro-CT, slice at 7 mm

(c) micro-CT, slice at 8 mm

(d) micro-CT, slice at 9 mm

(e) Tomosynthesis, slice at 7 mm

(f) Tomosynthesis, slice at 8 mm

(g) Tomosynthesis, slice at 9 mm

Figure 2.11: Reconstruction of a finger bone. Comparison of tomosynthesis and micro-CT slices, which are located at 7 mm, 8 mm and 9 mm height.

Table 2.3: Technical parameters of the Siemens Somatom Definition clinical CT device, the Skyscan 1172 Micro-CT device and the Siemens Mammomat Inspiration DT device.

Parameter	CT	Micro-CT	Tomosynthesis
Angular range θ / deg	360	360	±25
Angular step size $\Delta\theta$ / deg	0.3	0.3-0.9	2
Number of projections N_{proj}	> 1000	> 1000	25
Acquisition time / sec	0.1-1	> 3600	30
X-ray tube	small cone angle	cone-beam	half-cone
Trajectory	spiral	circle	arc
Voltage / kV	140	20-100	≤35
Filter	-	Al, Cu, Al+Cu, selected by user	Rhodium (0.05 mm)
Current / mA	100-500	20-100	100-500
Detector type	curved, multi-row	flat-panel	flat-panel
Detector size / mm	2400	54×22	240×300
Detector element size / mm	3	0.022	0.085
Detector size / pix	(736 × 2) × 16	1280×1280	3584×2816
Raw-data / GB	> 1	> 0.5	0.5

Table 2.4: Reconstruction parameters of the Siemens Somatom Definition clinical CT device, the Skyscan 1172 Micro-CT device and the Siemens Mammomat Inspiration DT device.

Parameter	CT	Micro-CT	Tomosynthesis
Object size / mm	human body	4 × 4 × 4	200 × 300 × 100
Number of images (slices)	> 1000	> 1000	30-100
Image size / pixels	512×512	1280×1280	3584×2816
Slice thicknes / mm	0.6	0.02	1
Voxel size / mm	0.6 × 0.6 × 0.6	0.02 × 0.02 × 0.02	0.085 × 0.085 × 1
Resolution	isotropic	isotropic	non-isotropic, limited in-depth
Image orientation	perpendicular to the detector	perpendicular to the detector	parallel to the detector
Reconstruction time / sec	real time	> 3600	60
Reconstruction / GB	0.1-1	1-10	> 0.8

2.5 Problems of limited data tomography

A tomosynthesis data acquisition results in an incomplete dataset. In this section several theoretical aspects of image reconstruction in case of limited data will be discussed.

The tomographic reconstruction problem can be formulated as follows. Let Ω denote the support of the object (the set of points over which it is nonzero). The non-negative function μ describes the X-ray attenuation coefficients and is zero outside Ω. If an ideal infinitely thin monoenergetic X-ray beam with initial intensity I_0 passes through the object along a straight line l, then the measured intensity I after passing the object will be

$$I = I_0 e^{-\int_l \mu(s) ds}. \tag{2.5}$$

If we define $f_L = -ln\,(I/I_0)$ and assume that L is the set of all lines in Ω, then

$$f_L = \int_L \mu(s)\, ds, \tag{2.6}$$

where s denotes a measure along L. The reconstruction problem is to recover the unknown function μ based on the set of line integrals f_L (Radon 1917, Cormack 1963, Cormack 1964).

2.5.1 Types of limited data problems

Four types of limited data problems exist, according to the classification by T. Quinto (Quinto 2012).

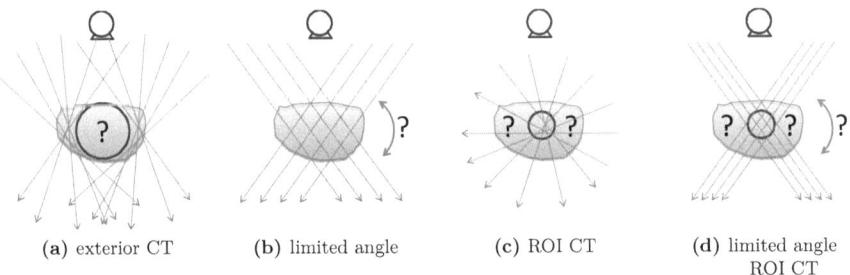

(a) exterior CT   (b) limited angle   (c) ROI CT   (d) limited angle ROI CT

Figure 2.12: Types of limited data according to T. Quinto (Quinto 2012).

The first type is called the *exterior CT* problem (Fig. 2.12a). Here, only the data outside an excluded region (marked by a circle) are measured. The task is to recover the object outside this region. The second type is called the *limited angle CT* (Fig. 2.12b). Only the data within a limited angular range are available. A unique solution exists but it is very unstable. The third type is called the *ROI CT*, where measurements are done within a limited region of interest (marked by a circle). Because of the overlapping principle of the CT measurements, also the contribution from the object outside the ROI is included into the measured data. Nevertheless, only the reconstruction of the selected region of interest is required (Fig. 2.12c). No unique solution exists (Noninjectivity Theorem). The fourth type is called the *limited angle ROI CT*. It is a combination of the second and the third type (Fig. 2.12d). The tomosynthesis problem can be classified as the second type of limited data problem.

2.5.2 Radon transform and singularities

The Radon transform is the most important transform in the mathematical theory of tomography. Johann Radon proposed a mathematical technique how to determine a function based on its line integrals in 1917 (Radon 1917). This paper was later translated to English (Radon 1986). Allan M. Cormack independently studied the same problem and published two papers in 1963 and 1964 (Cormack 1963, Cormack 1964). As he mentioned in the Nobel lecture (Cormack 1979), he learned only fourteen years later that Radon already had solved this problem in 1917.

Let a direction vector θ belong to a unit sphere $\theta \in S^{n-1}$ and $\Theta^\perp = \{t\theta^\perp : t \in \mathbb{R}\}$ be a hyperplane through the origin perpendicular to θ. The Radon transform R of a function $f \in L_1(\mathbb{R}^n)$ is defined by a line integral over a hyperplane which is perpendicular to the direction θ with signed distance s from the origin

$$Rf(\theta, s) = R_\theta f(s) = \int_{\Theta^\perp} f(x + s\theta) \, dx, \, s \in \mathbb{R}. \quad (2.7)$$

The X-ray transform P of a function $f \in L_1(\mathbb{R}^n)$ is defined by a line integral over a line $l(\theta, y)$ which is parallel to the direction θ and which passes through a point $y \in \Theta^\perp$

$$Pf(\theta, y) = P_\theta f(s) = \int_\mathbb{R} f(x + y\theta) \, dx, \, y \in \Theta^\perp. \quad (2.8)$$

In two dimensions, the Radon transform and the X-ray transform differ from each other only in the parameterization, i.e. they are both defined as a line integral. In three dimensions, the Radon transform is an integral over a plane and cannot be used to model tomographic acquisition. The X-ray transform is described as a line integral for any dimension, and, therefore, is used for modeling. For the properties of the Radon

2.5 Problems of limited data tomography

and X-ray transforms see e.g. the papers from A. Faridani (Faridani 2003) and E. T. Quinto (Quinto 2006).

The tomographic reconstruction problem in two dimensions is to find a good approximation of the function f based on Rf acquired over a unit sphere S^1. The tomosynthesis reconstruction problem is to reconstruct the function f based on Rf acquired on a restricted subset of a unit sphere S^1_θ. Uniqueness of the solution requires an infinite number of lines be measured. In practice, only a finite number of lines can be measured. Reconstruction from the limited angle data is more ill-posed than reconstruction from complete data (Quinto 1993). A unique solution exists but it is unstable. Only certain features of the object can be reconstructed.

A Singularity of the object is defined as a density jump between material μ_1 and μ_2 or a boundary between regions with different tissues, i.e. where the density function is not smooth. Following the works by T. Quinto (Quinto 1993, Quinto 2007) only some singularities can be stably reconstructed from the limited data. More specific, only those boundaries of the object can be reconstructed, which can be "seen" by the source. In other words, there must be an integral measured along a line, which is perpendicular to the singularity **n** in the current point, see Fig. 2.13a. Other singularities are called *invisible* and cannot be stably reconstructed, see Fig. 2.13b.

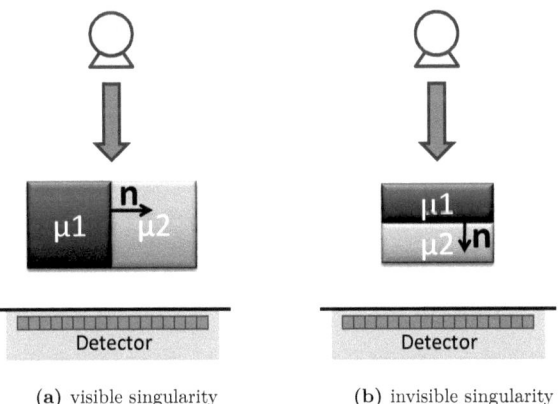

(a) visible singularity (b) invisible singularity

Figure 2.13: Illustration of a visible (a) and an invisible (b) singularities of an object.

Based on this principle, it can be demonstrated, which features are reconstructible in tomosynthesis. Let us consider a virtual two-dimensional phantom, which consists of two circles, see Fig. 2.14. The fan-beam X-ray tube is moved along an arc from the position A to the position B. At each X-ray tube position, it casts two *visible* points on each circle. Based on the acquisition measured in the position A, singularities of the left sphere in points A_1 and A_2 can be reconstructed, see Fig. 2.14a. Accordingly, singularities in points B_1 and B_2 can be restored based on the acquisition from the position B. The complete movement from A to B makes singularities on circle segments A_1B_1 and A_2B_2 visible. It is important to note, that on each circle a unique set of points is visible, compare A_1B_1, A_2B_2 and A_3B_3, A_4B_4.

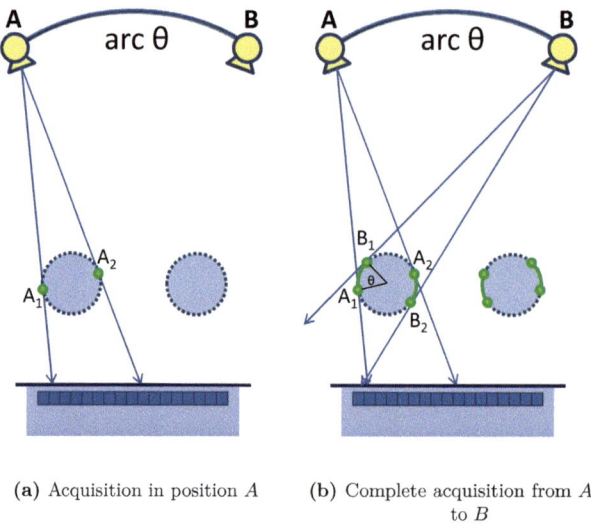

(a) Acquisition in position A (b) Complete acquisition from A to B

Figure 2.14: Illustration of visible and invisible singularities for a two-dimentional tomosynthesis fan-beam geometry. (a) The X-ray tube in position A makes singularities A_1 and A_2 visible. (b) The acquisition over the arc from A to B makes singularities A_1B_1 and A_2B_2 visible. Different parts of the left and the right spheres are visible.

In the three-dimensional case, the X-ray tube casts a *visible* ring on a sphere object. The orientation of visible rings depends on the position (x, y) and height z of the sphere.

2.5.3 Incomplete Fourier space

The Fourier transform is another major transform in the field of image reconstruction. The Fourier transform $\widehat{F}(\boldsymbol{\omega})$ of a real or a complex-valued n-dimensional function $f(\mathbf{x})$

2.5 Problems of limited data tomography

is defined as the integral

$$\widehat{F}(\boldsymbol{\omega}) = \left(\frac{1}{2\pi}\right)^{\frac{n}{2}} \int\limits_{R^n} f(\mathbf{x}) e^{-i\boldsymbol{\omega}\mathbf{x}} d\mathbf{x}, \qquad (2.9)$$

and its inverse is defined as

$$f(\mathbf{x}) = \left(\frac{1}{2\pi}\right)^{\frac{n}{2}} \int\limits_{R^n} \widehat{F}(\boldsymbol{\omega}) e^{i\boldsymbol{\omega}\mathbf{x}} d\boldsymbol{\omega}. \qquad (2.10)$$

An alternative definition of the Fourier transform exists, in which the coefficient 2π is inside the exponent in the integral. This is equivalent to the definition in terms of ordinary frequency u instead of angular frequency ω

$$\widehat{F}(\mathbf{u}) = \int\limits_{R^n} f(\mathbf{x}) e^{-2\pi i \mathbf{u}\mathbf{x}} d\mathbf{x}, \qquad (2.11)$$

and its inverse is

$$f(\mathbf{x}) = \int\limits_{R^n} \widehat{F}(\mathbf{u}) e^{2\pi i \mathbf{u}\mathbf{x}} d\mathbf{u}. \qquad (2.12)$$

It is also possible to exchange the signs in front of the exponent or to introduce a coefficient only in front of one equation instead of splitting it symmetrically. Different definitions in terms of frequencies are possible because of the scaling property of the Fourier transform which states that the Fourier transform of $f(\alpha x)$ is $(1/|\alpha|) \widehat{F}(u/\alpha)$. More information about the Fourier transform and its properties can be found e.g. in (Papoulis 1962).

The Fourier transform is the basis for the Fourier Slice Theorem (FST) (Bracewell 1956, Merserea 1974), which states that a one-dimensional Fourier transform of measured projection data $F_1\{p(\xi)_\theta\}$ under the direction view θ lies on a line, which crosses the origin of the two-dimensional Fourier transform of the image $F_2\{f(x,y)\}_\theta$

$$F_2\{f(x,y)\}_\theta = F_1\{p(\xi)_\theta\}. \qquad (2.13)$$

A schematic representation of the tomosynthesis measurement process and the filling of the Fourier space according to the FST are shown in Fig. 2.15. The tomosynthesis acquisition with an X-ray tube trajectory along an arc $\theta_x = \angle AOB$ covers only the limited wedge $\theta_x = \angle AOB = \angle A_\omega O_\omega B_\omega$ in the Fourier domain. The incompleteness of the Fourier domain results in artifacts in the reconstructed images, because information about features oriented along certain directions is unavailable.

2.5.4 Tuy-Smith sufficiency condition

Tuy and Smith independently in 1983 and 1985 derived a sufficiency condition for an object reconstruction (Tuy 1983, Smith 1985). The sufficiency condition states that one can reconstruct the object exactly if on every plane that intersects the support of the object Ω there exists at least one X-ray source point. Examples of trajectories which fulfill the condition are: two orthogonal circles, a spiral or two parallel circles connected by a line(Buzug 2008). A circular trajectory, which is usually used in micro-CT does not fulfill the conditions and neither does the tomosynthesis limited angle acquisition geometry. However, one can find some planes which fulfill the condition, compare Fig. 2.16.

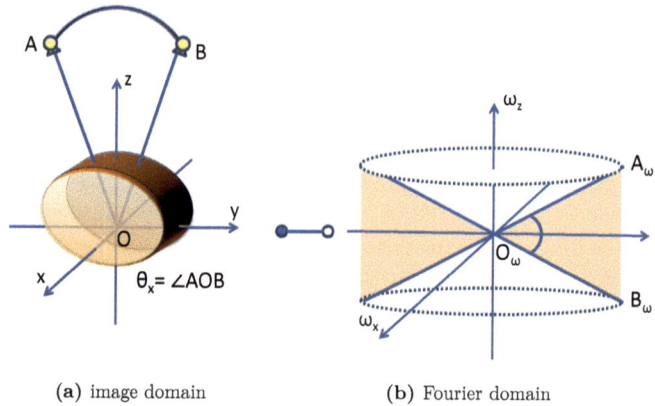

(a) image domain   (b) Fourier domain

Figure 2.15: Illustration of the incomplete Fourier domain. Only the limited wedge $\theta_x = \angle AOB = \angle A_\omega O_\omega B_\omega$ is filled during tomosynthesis acquisition along an arc trajectory.

2.5.5 Artifacts and limited resolution

Although tomosynthesis is a volumetric imaging technique and provides dimensional information about structures, the complete three-dimensional information about the object cannot be reconstructed. As it was discussed in the previous section, the incompleteness of the tomosynthesis projection data results in a missed wedge in the Fourier space and lost singularities of the Radon transform (Quinto 1993). Furthermore, the Tuy-Smith sufficiency condition (Tuy 1983, Smith 1985) are not fulfilled. Therefore,

2.5 Problems of limited data tomography

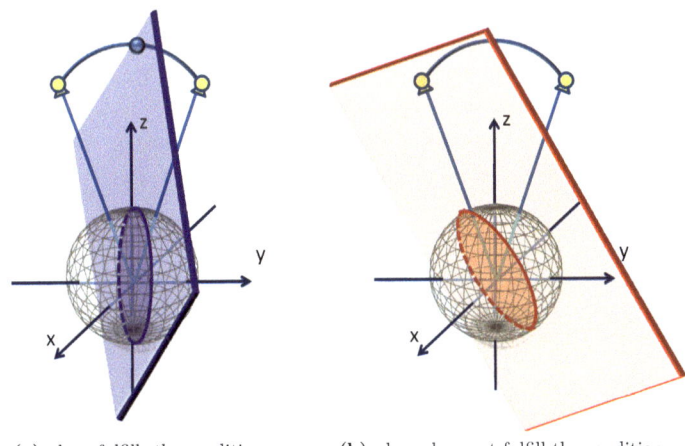

(a) plane fulfills the condition (b) plane does not fulfill the condition

Figure 2.16: Illustration of a plane which fulfills the Tuy-Smith sufficiency condition (a) and doesn't fullfill the condition (b) in tomosynthesis acquisition along an arc trajectory.

an accurate reconstruction in tomosynthesis is a challenging task. The incompleteness of the projection data results in artifacts and limits the axial resolution of the reconstructed volume. Thus, tomosynthesis yields images with an anisotropic resolution.

In the xy-planes artifacts corrupt the appearance of boundaries and might hide fine structures. Another appearance of limited angle artifacts is that shapes appear distorted in axial slices and features are surrounded by a sandglass-shaped halo in z-direction. It is known that the intensity and an artifact spread in z-direction is proportional to the size and density of the artifact-causing features (Svahn 2007, Reiser 2007, Hu 2008a).

One appearance of limited angle artifacts are out-of-focus artifacts (also called structural noise) which appear on the target plane and are produced by structures located above or below the current plane. The formation and propagation of such artifacts through the volume is shown in Fig. 2.17. The object is an apple with several metal needles inserted into it parallel to the detector with a 10 mm spacing. At a height of 48 mm two needles are inserted. On the two-dimensional projection image all six needles are visible, but the information about their location in z-direction is lost. The reconstructed results show that all needles are reconstructed on their correct positions, see the slice at 11 mm (Fig. 2.17c) and the slice at 48 mm (Fig. 2.17e). At the same time, strong out-of-focus artifacts are presented in the slice at 25 mm (Fig. 2.17e) as

multiple ghosting copies of needles from slices above and below.
An example of out-of-focus artifacts in clinical images of a hand is shown in Fig. 2.18b. These artifacts are highlighted by arrows. The structures in the axial slices have a distorted shape and are surrounded by a triangle-shaped halo (Fig. 2.18c).

Limited angle artifacts limit the diagnostic value of DT images. Therefore a lot of effort is made to investigate how to improve tomosynthesis performance and obtain images with better quality.

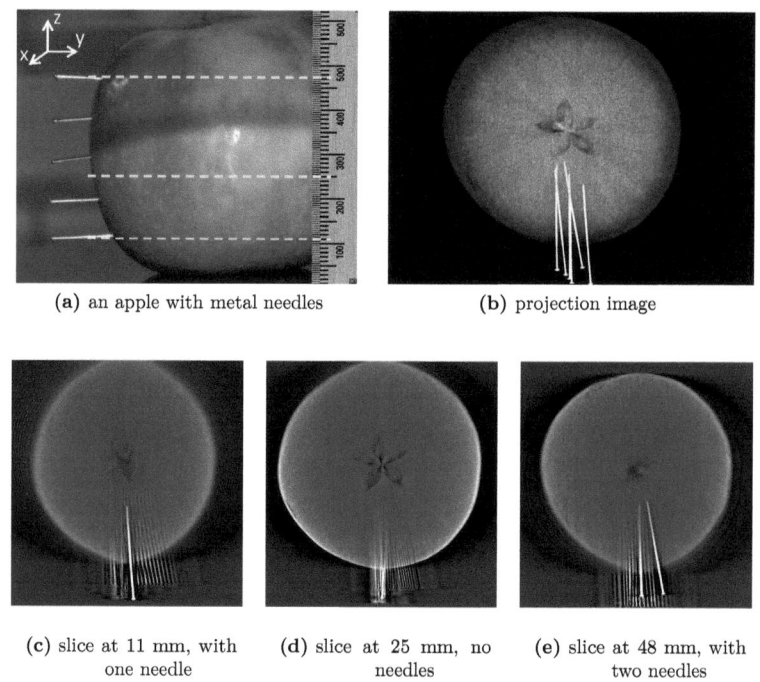

(a) an apple with metal needles   (b) projection image

(c) slice at 11 mm, with one needle   (d) slice at 25 mm, no needles   (e) slice at 48 mm, with two needles

Figure 2.17: Tomosynthesis raw-data and reconstructed slices (Siemens FBP) of an apple with five needles, 10 mm spacing. Images illustrate a propagation of the out-of-focus artifacts produced by high-attenuation objects (needles).

2.5 Problems of limited data tomography 41

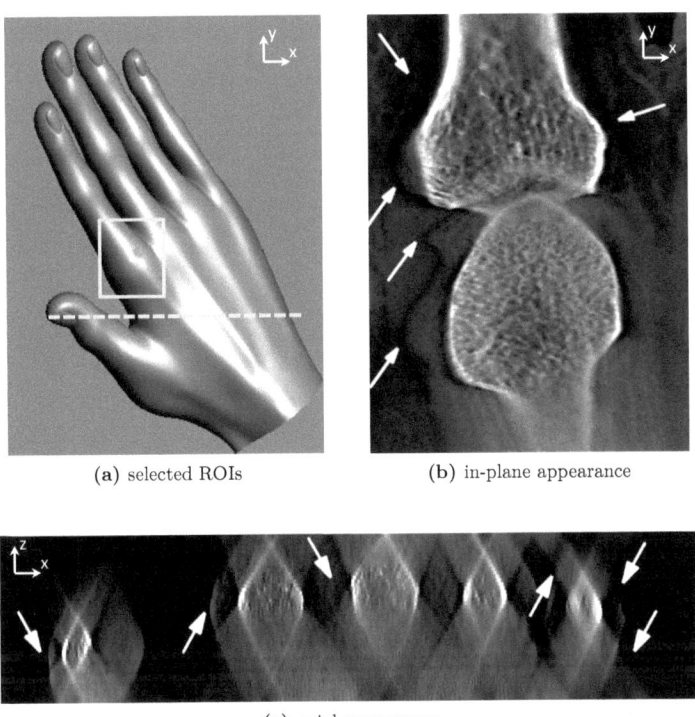

(a) selected ROIs (b) in-plane appearance

(c) axial appearance

Figure 2.18: The reconstructed volume of a hand shows artifacts in tomosynthesis. (b) in-plane appearance; (c) axial appearance

Chapter 3

Forward and backprojection model (FP/BP)

Contents

3.1	Discrete image representation using series expansion	44
3.2	Pixel basis functions	45
3.3	FP/BP algorithms for pixel basis functions	46
3.4	Kaiser-Bessel basis functions (blobs)	49
3.5	FP/BP algorithms for blob basis functions	55
3.6	Efficient distance-driven projector in 2D	58
3.7	Efficient distance-driven projector in 3D	69

The forward projection (FP) operation is the mathematical model of the physical data acquisition process in computed tomography and tomosynthesis. The backprojection (BP) operation is the corresponding reverse model. The forward- and backprojection pair is a key module in reconstruction algorithms. In any iterative reconstruction algorithm this pair is used during each iteration at least once, therefore an optimal practical implementation is required to be fast, accurate and memory efficient. For the practical numerical implementation of the forward and backprojections, a continuous object has to be discretized, i.e. it has to be represented by a finite number of parameters.

This chapter addresses three problems: how to model the discrete representation of the continuous object, how to model the forward and backprojections operation and how to practically implement it efficiently and accurately. First, an image representation using square non-overlapping pixels and overlapping spherically symmetric Kaiser-Bessel functions (blobs) will be discussed in this chapter. Second, a literature review of existing forward and backprojection models for both types of basis functions will be presented. Finally, a practical implementation strategy for a state-of-the-art algorithm called distance-driven will be given. First, the implementation will be discussed for two-dimensional fan-beam CT geometry for pixels and blobs. Then, the algorithm for pixels will be extended to the three-dimensional cone-beam CT geometry and adapted for the tomosynthesis with the fixed detector.

3.1 Discrete image representation using series expansion

For a numerical implementation of the iterative reconstruction algorithms a continuous object has to be discretized, i.e. represented as a finite set of points. The choice of the representation model is important because it influences the accuracy of the reconstructed image and the speed and the computational complexity of the algorithm. For example, the set of points f_i can be a set of sampling points. A value at any arbitrary position is calculated using an interpolation operator. Alternatively, the set of points can be a set of series expansion coefficients (Hanson 1985). In many works on CT the discrete image representation is not discussed. In fact, it has been overseen that the choice of the discrete image representation has its own contribution to the reconstructed image quality. The focus of this chapter is the series expansion approach using basis function.

Let $\hat{f}(\mathbf{x})$ be a discrete approximation of a continuous three-dimensional distribution of the X-ray attenuation coefficients $f(\mathbf{x})$. The discrete approximation is defined at the spatial positions $\mathbf{x}$. It can be represented as a linear combination of scaled and shifted copies of some basis functions

$$f(\mathbf{x}) = \hat{f}(\mathbf{x}) \approx \sum_{i=1}^{N} c_i b_i (\mathbf{x} - \mathbf{x_l}). \tag{3.1}$$

Here, $\{c_i\}_{i=1}^{N}$ is a set of expansion coefficients, $\{b(\mathbf{x})\}_{i=1}^{N}$ is set of known basis functions, arranged on the a three-dimensional grid with N total number of grid points $\mathbf{x_l}$. It is important to note that especially in case of overlapping basis functions, the expansion coefficients c are not equal to the sampling values of the function f.

Based on a classification with respect to the spatial support, two different kinds of basis functions exist. Basis functions with an unlimited support are called *global* basis functions, for example exponential basis functions of the Fourier series expansion. Another type of basis functions are *local* basis functions. They are localized in space and are zero-valued outside a limited support. The focus of the further discussion will be only on the local basis functions. Several different choices of the local basis functions exist. This choice will define the complexity of the involved mathematical calculations and, therefore, the speed of the algorithm. Contrary, a simple model could be fast but it might lack high accuracy.

A traditional choice for basis functions are non-overlapping *pixels* or *voxels* for the two-dimensional and the three-dimensional case, respectively. A pixel basis function has a value equals one inside a square and zero outside. Another choice may be generalized Kaiser-Bessel functions, also known as *blobs* (Lewitt 1992). They are localized in space as are voxels. A smooth bell-shaped radial profile and the overlapping nature of blobs allows for creating a smooth representation of naturally smooth biological objects. The X-ray integrals of blobs can be calculated analytically. The spherical symmetry of blobs and their X-ray transforms allows for efficient calculation of line integrals. At the same time, the overlapping degree increases the computational complexity.

R. M. Lewitt (Lewitt 1992) mentioned that the spherical Gaussian function have also a number of attractive properties: its Fourier transform and its projection are Gaussian. However, this function is not completely localized in space and must be truncated for the practical implementations. The truncated version of the Gaussian function does not have abovementioned properties anymore. Another alternative choice of the basis function may be b-splines (Entezari 2012). For the comparison of the bias introduced by blobs and b-spline basis functions see (Schmitt 2012). In the rest of the chapter the discussion will be limited by considering only the pixels and the blob basis functions.

Not only the choice of the basis functions is important. It is also important to choose the model how to evaluate the basis functions in forward- and backprojection operations. A realistic modeling provides better image quality but, at the same time, it increases the complexity of algorithm. Both, the choice of the basis function type and the choice of the evaluation method is always a compromise between the accuracy and complexity of practical implementation.

3.2 Pixel basis functions

Let a grid be a three-dimensional regular uniform Cartesian grid (simple cubic grid) with a grid increment Δ. A *pixel* basis function is non-zero only in the local region, described by the corresponding spatial location. It can be described as

$$b(\mathbf{x}) = \begin{cases} 1, |\mathbf{x}| < \Delta/2 \\ 0, \text{otherwise} \end{cases}. \tag{3.2}$$

A profile of the basis function is shown in Fig. 3.1a. A function approximated by pixels is piece-wise constant, see Fig. 3.1b. In general, an image approximation by square pixels in terms of series expansion is equivalent to the nearest neighborhood interpolation.

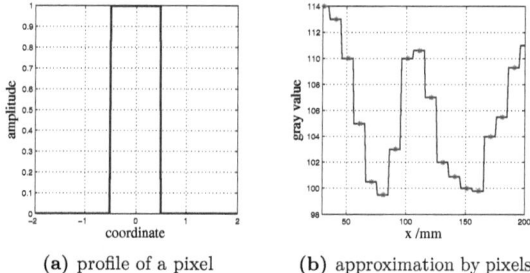

(a) profile of a pixel   (b) approximation by pixels

Figure 3.1: Pixel basis function. (a) a profile of a pixel; (b) a signal approximation using pixels.

An analytical formula of the Radon transform of the pixel basis function is given by

$$\begin{cases} g(\rho,\theta) = 0, \ x_1 > 0 \\ g(\rho,\theta) = \sqrt{2^2 + (x_1 - x_{-1})^2} = \frac{2}{\cos\theta}, \ x_1 < 1 \text{ and } x_{-1} < 1 \\ g(\rho,\theta) = \sqrt{(1-x_1)^2 + (1-x_{-1})^2}, \ x_1 < 1 \text{ and } x_{-1} > 1 \end{cases} \quad (3.3)$$

where x_1 and x_{-1} are intersections between boundaries $y = -1, y = 1$ with the line $p = x\cos\theta - y\sin\theta$. The detailed derivation of this formula can be found in the appendix B of (Toft 1996). The pixel basis function is the easiest and most popular choice for basis functions for the image representation and modeling of the forward and backprojection operators in many tomographic imaging techniques.

3.3   FP/BP algorithms for pixel basis functions

The importance of a fast and efficient implementation of the forward- and backprojection algorithms was recognized quite soon after the establishment of the computed tomography. One of the first projector routines mentioned in the literature can be found in the RECLBL Library Package (Huesmanand 1977), namely the *pixel-driven* and the *ray-driven* approaches. These two approaches have mostly remained as the basis for all subsequent versions.

3.3 FP/BP algorithms for pixel basis functions

The pixel-driven principle is the simplest algorithm (Peters 1981). Each pixel is visited in a loop and the center of each pixel is projected on the detector according to the geometry, see Fig. 3.2a. For the pixel-driven forward projection a contribution of the pixel is split between two neighboring detector elements using typically a linear interpolation, or more complex interpolation schemes, e.g. bicubic spline interpolation (Harauz 1983). Fessler et al. (Fessler 1997a) proved the equivalence of the pixel-driven and a rotation-based approaches. This paper was rejected by IEEE Transactions on Medical Imaging as being "too obvious". The pixel-driven forward projection is rarely used because it tends to introduce high-frequency artifacts to the sinogram. If the detector element size is much smaller than the pixel element size, there is a danger to have detector elements in which no value is written. In a similar manner, in the pixel-driven backprojection, each pixel element is updated based on the value which is obtained from the neighboring detector elements, typically using linear interpolation. The pixel-driven backprojection is preferred for FBP reconstruction. The variations of the pixel-driven method exist, e.g. when each pixel is divided into four sub-pixels for higher accuracy. Alternatively, a technique called *splatting* has been developed to reduce aliasing artifacts by means of casting a smooth footprint for each pixel (Mueller 1998b, Birkfellner 005). Typically, all modifications introduced into the original algorithm result in higher computational costs.

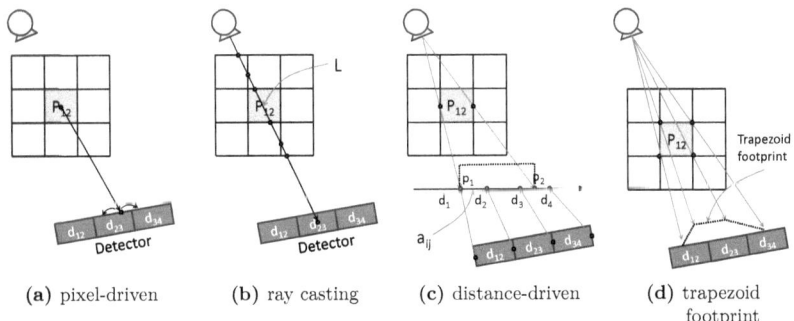

Figure 3.2: Models of forward and backprojectors for pixel basis functions. (a) pixel-driven; (b) ray casting; (c) distance-driven; (d) trapezoid footprint.

In the ray-driven (ray casting) approach the X-ray source and the detector elements are connected by straight lines according to the acquisition geometry, see Fig. 3.2b. The projection value is calculated as a weighted sum of all pixels that lie on the line. The backprojection operator can be defined as the transpose of the forward projection operator. A value from the detector element is back-distributed along the line using some weighting. The exact type of the weighting varies with different methods, but mainly intersection length between each pixel and the ray is considered, see e.g. (Gullberg 1985). The ray-driven backprojection typically results in high-frequency artifacts in the image (the Moire pattern). Two classical ray-driven approaches are Joseph's algorithm (Joseph 1982) and Siddon's algorithm (Siddon 1985). P. M. Joseph proposed to use the linear interpolation between pixels in the ray to increase the accuracy of the projectors. R. L. Siddon developed the exact fast algorithm for calculating the path length of the ray through each pixel by considering that the volume can be described as an intersection of orthogonal sets of parallel planes. Recent works improve the Siddon's method by accelerating it (Jacobs 1998, Christiaens 1999, Zhao 2003, Gao 2012).

A variant of the ray casting algorithm is a strip-integral based method. The ray is assumed to have some width which is typically equal to the detector element size. The contribution from a pixel to the detector element is equal to an integral over the strip-shaped intersection area between the beam and the pixel, see e.g. (Lo 1988). The authors proposed to use a triangle subtraction method for the calculation of the strip integral.

Iterative reconstruction methods require that the forward and backprojection methods are the transposed of each other. This is called a *matched* projector pair. However, both methods, the pixel-driven and the ray-driven, introduce artifacts either in the sinogram or in the image domain. Those artifacts might be amplified through each iteration. This leads to an idea to combine the pixel-driven backprojector and ray-driven projector resulting in an *unmatched* projector pair. It was shown that in some cases the unmatched pair has better performance than the matched pair. For the analysis of the effects of usage the unmatched pairs is used see e.g. (Zeng 2000, Guedouar 2010).

Alternative projectors include Fourier-based methods (Tabei 1992), the "natural" pixel decomposition (Buonocore 1981, Bevilacqua 2007), projectors with different interpolation methods (Xu 2006) and the usage of the projection matrices (Galigekere 2003).

A fundamentally different approach called *distance-driven* was proposed by B. De Man for two dimensions in 2002 (De Man 2002, De Man 2003) and was afterwards extended to the third dimension (De Man 2004). In this approach the pixel footprint is approximated by a rectangle, see Fig. 3.2c. The pixel boundaries and the detector boundaries are considered in this approach instead of the pixel and the detector centers. The boundaries are projected onto a common axis according to the acquisition geometry and the overlap length is used as a weighting coefficient. The distance-driven approach has low arithmetic costs and avoids high-frequency artifacts in both, the forward- and backprojectors. The distance-driven approach has the most inaccuracy at 45° because the pixel footprint is rather a triangle than a rectangle at this view. In order to take into account a more accurate pixel footprint shape, Long et al. (Long 2010a, Long 2010b, Long 2011) proposed an algorithm based on the separable trapezoid footprints, Fig. 3.2d. A GPU acceleration of this projector can be found in (Wu 2011). The distance-driven and the separable trapezoid footprints algorithms are two state-of-the art algorithms.

3.4 Kaiser-Bessel basis functions (blobs)

The Kaiser-Bessel basis functions (blobs) are known as an attractive alternative to the pixel basis functions. They have been introduced into the field of medical imaging by R. M. Lewitt (Lewitt 1990). They are used in electron microscopy (Marabini 1998, Garduno 2004), positron emission tomography (PET) (Chlewicki 2004, Jacobs 1999b), single-photon emission tomography (SPECT) (Yendiki 2004, Wang 2004) and transmission tomographic techniques such as CT (Jacobs 1999a, Carvalho 2003, Zbijewski 2006, Isola 2008) and breast tomosynthesis (Wu 2010). Alternatively, one can find an application of a blob-shaped window function for the post-backprojection filtering to improve the pixel-based reconstruction (Zhang 2006a). One of the latest application of blobs is a derivation of the differential forward operator for the phase contrast imaging (Köhler 2011).

3.4.1 Properties of the blobs

The generalized Kaiser-Bessel basis function is a spherically symmetric basis function with a local support, see Fig. 3.3. It is defined as

$$b(r)_{m,\alpha,a} = \begin{cases} \frac{I_m\left(\alpha\sqrt{1-\left(\frac{r}{a}\right)^2}\right)}{I_m(\alpha)} \left(\sqrt{1-\left(\frac{r}{a}\right)^2}\right)^m, & 0 \leq r \leq a \\ 0, & \text{otherwise} \end{cases} \quad , \quad (3.4)$$

where I_m is a modified Kaiser-Bessel function of the order m and r is the radial distance from the center of the blob. An image approximated by blobs is smooth, see Fig. 3.3b. The taper parameter a defines the support size (radius) of the blob, see Fig. 3.4a. The blob radius is measured in the grid size step units. Blobs with larger radius have an overlap with more neighboring blobs. The larger the area of overlap, the larger the computational demand. The parameter α controls the shape of the blob bell. The smaller the value of this parameter, the wider the shape of the blob, see Fig. 3.4b. The

Chapter 3. Forward and backprojections

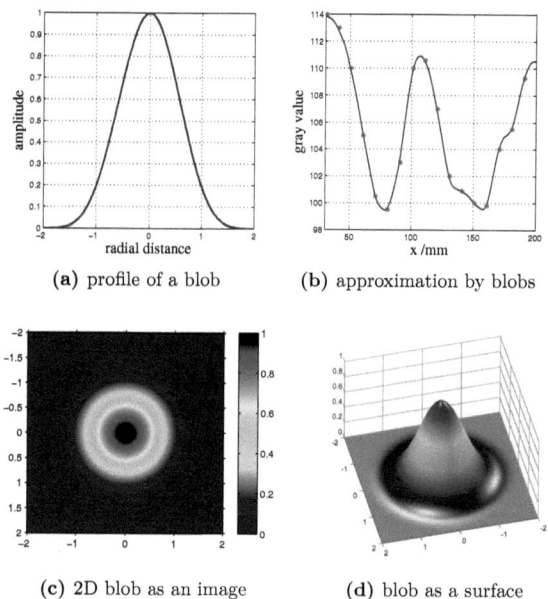

Figure 3.3: Different representations of a blob basis function with parameters $a = 2$, $m = 2$, $\alpha = 10.8$. (a) a profile of the blob; (b) a signal approximation approximation using blobs; (c) representation of the blob as an image; (d) representation of the blob as a surface.

order of the Kaiser-Bessel function m defines the continuity of the resulting image and its derivatives. It has also an influence on the shape of the bell, see Fig. 3.4c. But this effect is less significant than the influence of the parameter α.

For the special case of the rotationally symmetric function b_n^m, the n-dimensional Fourier transform $\hat{b}_n^m$ can be expressed as the Hankel relation (Lewitt 1990)

$$\hat{b}_n^m(R) = \frac{2\pi}{R^{n/2-1}} \int_0^\infty b_n^m(r) J_{n/2-1}(2\pi R r) r^{n/2} dr. \tag{3.5}$$

A substitution of the integration variable from θ to $r = a \cos \theta$ and taking into account the relation between Bessel functions

$$I_v(z) = e^{-iv2\pi/2} J_v(iz) \tag{3.6}$$

lead to an integral which, according to R. M. Lewitt, has a form of "Sonine's second

3.4 Kaiser-Bessel basis functions (blobs)

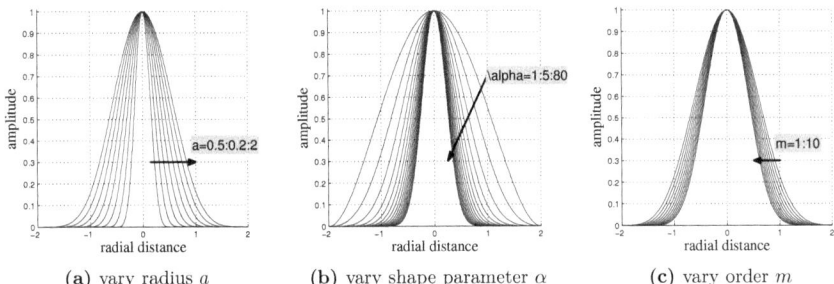

Figure 3.4: Influence of the blob parameters on the shape and size of the blob. (a) a controls the radius; (b) α controls the bell wideness; (c) m controls the continuity order of the image and the bell wideness.

finite integral"

$$\hat{b}_n^m(R) = \frac{2\pi i^{-m}}{I_m(\alpha)} \frac{a^{n/2+1}}{R^{n/2-1}} \int_0^{\pi/2} J_m(i\alpha \sin\theta) J_{n/2-1}(2\pi aR\cos\theta) \cos\theta^{n/2} \sin\theta^{m+1} d\theta. \quad (3.7)$$

The resulting expression of the Fourier transform obtained from this integral is

$$\hat{b}_n^m(R) = \begin{cases} \frac{(2\pi)^{n/2} a^n \alpha^m}{I_m(\alpha)} \frac{I_{n/2+m}\left[\sqrt{\alpha^2-(2\pi aR)^2}\right]^{1/2}}{\left\{[\alpha^2-(2\pi aR)^2]^{1/2}\right\}^{n/2+m}}, & 2\pi aR \leq \alpha \\ \frac{(2\pi)^{n/2} a^n \alpha^m}{I_m(\alpha)} \frac{J_{n/2+m}\left[(2\pi aR)^2-\alpha^2\right]^{1/2}}{\left\{[(2\pi aR)^2-\alpha^2]^{1/2}\right\}^{n/2+m}}, & 2\pi aR \geq \alpha. \end{cases} \quad (3.8)$$

The obtained Fourier transform of the blob basis functions is rotationally symmetric. The parameter m controls the decay rate and the parameter α controls the amplitude of the Fourier transform. The one-dimensional case with parameters $m = 0$, $n = 1$, $\alpha = 0$ simplifies the blob function to the well-known $rect$ function. The Fourier transform (equation 3.8) derived in terms of the definition of the blob function (equation 3.4) is simplified to the well-known Fourier transform of the $rect$ function

$$\hat{b}_{n=1}^{m=0}(R) \propto \frac{\sin x}{x} \quad (3.9)$$

if the following relation

$$J_{1/2}(z) = \sqrt{\frac{2}{\pi z}} \sin z \quad (3.10)$$

is taken into account (Lewitt 1990).

The function is called effectively band limited within a spectral radius A at the level ε if the following equality holds (Lewitt 1990, Jacobs 1999a)

$$\frac{\int_{\|R\|<A} \left\|\hat{b}(R)\right\|^2 dR}{\int_{\|R\|<\infty} \left\|\hat{b}(R)\right\|^2 dR} = 1 - \varepsilon. \tag{3.11}$$

The two-dimensional Fourier transform of a two-dimensional blob function with parameters $m = 2$, $a = 2$, $\alpha = 10.83$ is shown in Fig. 3.5a at a logarithmic scale. A radial line profile of the Fourier transform is shown in Fig. 3.5b. The Fourier transform

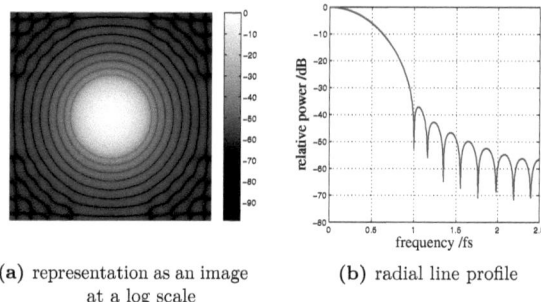

(a) representation as an image at a log scale

(b) radial line profile

Figure 3.5: Two-dimensional Fourier transform of a two-dimensional blob function with parameters: $m = 2$, $a = 2$, $\alpha = 10.83$. (a) representation as an image at a logarithmic scale; (b) the radial line profile

of this blob ($m = 2$, $a = 2$, $\alpha = 10.83$) is effectively band limited within the spectral radius of $A = 1$ at the level $\varepsilon = 0.008$.

The X-ray transform of a blob basis function is given by

$$p(s) = 2 \int_0^{(a^2-s^2)^{1/2}} b_n^m \left[\left(s^2 - t^2\right)^{1/2}\right] dt. \tag{3.12}$$

R. M. Lewitt (Lewitt 1990) shows that the resulting expression is

$$p(s) = \frac{a}{I_m(\alpha)} \left(\frac{2\pi}{\alpha}\right)^{1/2} \left[\sqrt{1 - (s/a)^2}\right]^{m+1/2} I_{m+1/2}\left[\alpha\sqrt{1 - (s/a)^2}\right], \tag{3.13}$$

where s is a distance from the blob center to the X-ray and $\sqrt{a^2 - s^2}$ is one half of the intersection length between the blob and the ray, see Fig. 3.6a. This is typically

3.4 Kaiser-Bessel basis functions (blobs)

called the *footprint* of a blob. The projection value through the blob depends only on the distance s and does not depend on the angular direction. A radial profile of the X-ray transform of the blob basis functions for $m = 2$, $a = 2$ and varying parameter α is shown in Fig. 3.6b. Further discussion on the blob basis functions can be found in (Lewitt 1990).

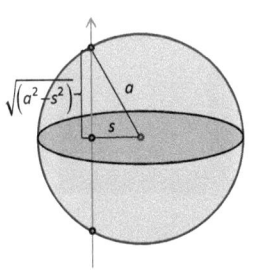
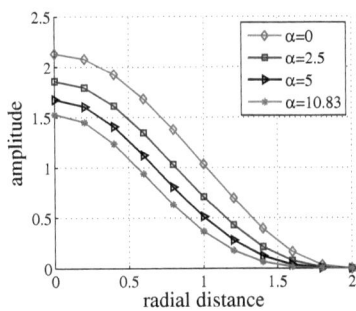

(a) intersection of a ray and a blob   (b) radial profile of the X-ray transform

Figure 3.6: X-ray transform of the blob basis function. (a) a schematic representation of an intersection of an X-ray and a blob; (b) a radial profile of the X-ray transform (projections) of the blob basis functions for $m = 2$, $a = 2$ and varying parameter α.

3.4.2 Finding the optimal parameters a, α, m

The shape of the blob must be optimized for the practical implementation. One optimization strategy is based on the constraint function representation (Matej 1996). It is logically to assume, that the blob shape should be selected in a way to represent a constant function with the minimum error

$$f(x, y) = \text{const} = \sum_{i=1}^{N} c_i b_i (x - x_i, y - y_i). \tag{3.14}$$

Let a uniform function $f(x, y)$ be discretized using a two-dimensional grid of 30×30 nodes with a node spacing $\Delta_{image}=1$. Let us construct a series expansion representation of this function using blob basis functions. Typically, it is desired to have a continuous image with the continuous derivative, therefore the parameter $n = 2$ is chosen. The radius of the blob defines the overlap area with neighboring blobs and is selected in a way, that the FWHM of the blob is not larger than the data resolution. For the numerical implementation, blobs also have to be discretized. Let blobs be discretized on a finer grid with node spacing $\Delta_{blob} = 0.05$. The superposition of too narrow blobs ($m = 2$, $a = 2$, $\alpha = 15$) results in an image with gaps, see Fig. 3.7a. Too wide blobs ($m = 2$, $a = 2$, $\alpha = 3$) also result in oscillations in the image, see Fig. 3.7c. At the same time, if parameters are optimal $m = 2$, $a = 2$, $\alpha = 10.8$, the representation of a uniform function is close to a constant value, see Fig. 3.7b.

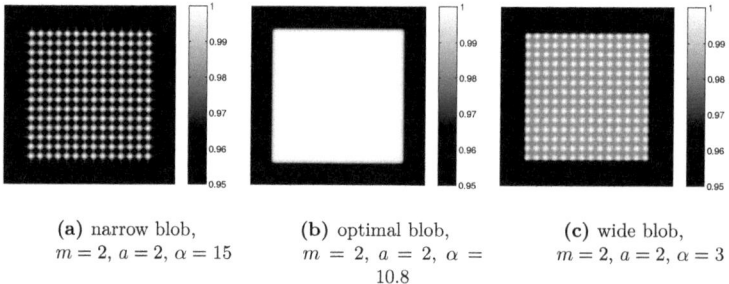

(a) narrow blob, $m = 2$, $a = 2$, $\alpha = 15$

(b) optimal blob, $m = 2$, $a = 2$, $\alpha = 10.8$

(c) wide blob, $m = 2$, $a = 2$, $\alpha = 3$

Figure 3.7: Constant function representation using (a) a narrow blow $m = 2$, $a = 2$, $\alpha = 15$; (b) an optimal blob $m = 2$, $a = 2$, $\alpha = 10.8$; (c) a wide blob $m = 2$, $a = 2$, $\alpha = 3$

The contour representation of the root mean squared error in percent between the constant-valued function and its approximation is shown in Fig. 3.8. The blobs with $m=2$ and selected pairs (α, a) were used as basis functions. Based on this plot, the optimal blob taper parameters for the radius $a = 2$ is $\alpha = 10.8$.

Another strategy to find optimal parameters for blobs is based on the Fourier transform (Matej 1996). It is known, that the Fourier transform of a constraint-valued function is an impulse at the origin. If Δ_{image} is the lattice interval of the periodic grid, then the Fourier transform of the basis function must be zero for the radial frequency $R = 1/\Delta_{image}$. Apparently, the blob with parameters found based on the constant function representation fulfills this condition, see Fig. 3.5b.

Another problem which must be addressed is the optimal grid layout for blobs. The Cartesian grid, which is natural for cubic voxels, is not natural for blobs. The different distances between the nodes in the horizontal/vertical and the diagonal directions make this grid inefficient in case of blobs. The alternative body-centered and face-centered cubic grids might be better suitable for image representation using blobs because they have a better packing efficiency[1] and a smaller number of blobs is required

---

[1] regarding the sphere packing the reader might be referred to the solid-state physics topic

3.5 FP/BP algorithms for blob basis functions

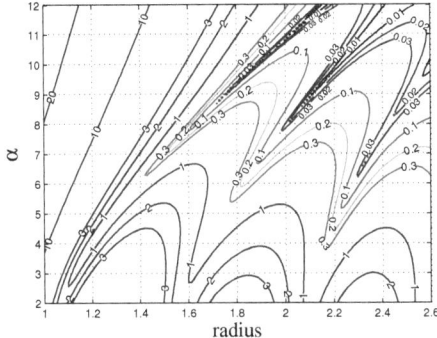

Figure 3.8: The contour representation of the root mean squared error in percent between the constant-valued function and its approximation using the blob ($m=2$) as basis functions. The optimal blob taper parameters for the radius $a = 2$ is $\alpha = 10.8$.

(Lewitt 1992, Garduno 2004). In general, the question of efficient packing of spheres, called Kepler conjecture (Marchal 2011) is still unproved and in dimensions bigger than three is still unsolved.

3.5 FP/BP algorithms for blob basis functions

3.5.1 Ray tracing through grid of blobs

A first design of the forward/backprojection operation for blobs was proposed in 1996 by S. Matej and R. M. Lewitt (Matej 1996). The authors discussed two methods, the ray-driven and the blob-driven. Later, Popescu et. al (Popescu 2004) proposed two different methods for ray tracing through a grid of blobs: without direct tracking of the distance along the ray and with tracking of the distance along the ray.

The ray-driven forward- and backprojector for blobs are also based on drawing a beam from the X-ray source to each detector element. The difference to the ray-driven algorithms for pixel basis functions is that not the intersection length is of interest but the distance between the ray and the center of each visited blob. Because of the overlapping property, more blob elements will contribute to the selected ray compared to the number of pixel elements in the similar case. All blobs which belong to a cylinder centered around the beam will contribute to that beam (Matej 1996), see Fig. 3.9.

To use a lookup table of X-ray integrals, a fast algorithm to compute the distance s between the ray and each visited blob is required. The algorithm without direct

Chapter 3. Forward and backprojections

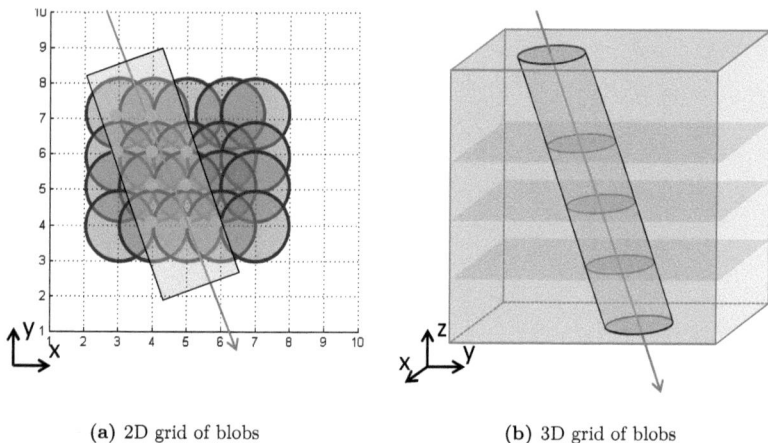

(a) 2D grid of blobs

(b) 3D grid of blobs

Figure 3.9: Ray tracing through grid of blobs in (a) 2D and (b) 3D. Blobs which contribute to the beam are lying within a rectangle in 2D and a cylinder in 3D.

tracking of the distance works as follows. If the ray direction is described by a point $\mathbf{a} = [a_1, a_2, a_3]$ and the direction $\mathbf{u} = [u_1, u_2, u_3]$, $|u| = 1$, then the squared distance r^2 to any blob center located at $x_{i1,i2,i3}$ is defined as a squared cross-product

$$r^2 = [(\mathbf{x} - \mathbf{a}), \times \mathbf{u}]^2 = \begin{vmatrix} \mathbf{e_1} & \mathbf{e_2} & \mathbf{e_3} \\ x_1 - a_1 & x_2 - a_2 & x_3 - a_3 \\ u_1 & u_2 & u_3 \end{vmatrix}^2 = \qquad (3.15)$$
$$r_{23}^2 + r_{32}^2 + r_{12}^2,$$

where

$$r_{i1,i2} = (x_{i_1} - a_{i_1})u_{i_2} - (x_{i_2} - a_{i_2})u_{i_1} = d_{i_1 i_2} - d_{i_2 i_1} \qquad (3.16)$$

with

$$d_{i_1 i_2} = (x_{i_1} - a_{i_1})u_{i_2}. \qquad (3.17)$$

Each grid point $\mathbf{x}$ can be described as $x = x_0 + \Delta x_i$. Then $d_{i_1 i_2}$ has two components: a constant component $(x_{01} - a_1)u_2$ and the incremental component $\Delta x u_2$.

Alternatively, based on the distance-driven algorithm for pixels, one can create a similar algorithm for blobs, resulting in an efficient implementation of strip-integral based algorithms (Levakhina 2010, Bippus 2011). The three-dimensional version of the blob-based forward and backprojection operators for blobs in the divergent cone-beam

3.5 FP/BP algorithms for blob basis functions

geometry using separable footprints was proposed by Ziegler et al. (Ziegler 2006). In the blob-driven approach, each blob is processed sequentially and the projections values are updated by adding the contribution from each corresponding blob (all projections lines which intersect the blob). Nowadays it is the state-of-the-art algorithm. Recently, several GPU accelerated implementations of the projectors for blob basis functions have been proposed (Wang 2010, Bippus 2011).

3.5.2 Lookup table calculation

A line integral through the blob basis function depends only on the distance s from the blob center to the integration line and has no dependency on the projection angle, see equation 3.13. The attractive spherical symmetry property allows for the pre-calculation of X-ray integrals and storing $p(s)$ as a look-up table (LUT) on a pre-selected fine grid of distances $s \in [0, a]$. For the ray-driven approach a LUT of line X-ray integrals is needed. For the strip-integral based approach (distance-driven approach) the LUT of areas of x-ray integral is needed. In the practical case, the line integral can be calculated using the analytical expression given in equation 3.13. Alternatively, it can be numerically approximated using equation 3.12. This can be done in MATLAB® e.g. using an adaptive Simpson quadrature which is implemented as a build-in function quad, see Listing 3.1. The LUT for area integrals can be easily calculated using the LUT of line integrals and the area rectangle method, trapezoidal rule, Simpson's rule or any other method for numerical integration. Using area integrals $p_{area}(t)$, where parameter t defines all possible distances to the blob center

$$p_{area}(t) = \int_{-\infty}^{t} p(s)ds, \; t = -(a, a) \tag{3.18}$$

a particular strip integral p_{strip} from t_1 to t_2 can be calculated as a difference of two corresponding area integrals

$$p_{strip} = \int_{-t_1}^{t_2} p(s)ds = p_{area}(t_2) - p_{area}(t_1). \tag{3.19}$$

Listing 3.1: MATLAB® code for an approximation of the x-ray integral through the blob using quad function

```
function [xray_integral] = do_xray_integral(m, a, alpha_blob, s)
% this function evaluates a line integral through a blob basis function at a ...
    distance s from the blob center.
quad_blob = 2 * quad(@define_blob,0,sqrt(a^2-s^2));
    function y=define_blob(t)  % define a blob formula
    Im1 = besseli(m,alpha_blob);
    r=sqrt(s.^2+t.^2);
    arg = sqrt(1-(r/a).^2);
    Im2 = besseli(m,alpha_blob*arg);
    y =Im2./Im1.*((arg).^m);
    end
end
```

3.6 Efficient distance-driven projector in 2D

In this section an efficient implementation of the distance-driven projector in two-dimensions will be described. The implementation is designed for MATLAB®.

3.6.1 General strategy

For simplicity, the implementation is described in the image coordinate system and not in the world coordinate system. All coordinates and lengths are given in image pixels and not in millimeters. The recalculation between the world coordinate system and the image coordinate system are trivial and not discussed here. Only the implementation of the forward projection will be given. The backprojection is defined to be adjoint[1] to the forward projection, i.e. it uses the same weighting coefficients.

The image is located in the first quadrant of the Cartesian coordinate system. The center of the left bottom pixel is located at $(1,1)^2$ and the bottom left pixel boundary (corner) is located at $(0.5, 0.5)$. It is assumed that the geometry of the detector is known and the coordinates of the detector boundaries are pre-calculated and saved as two vectors. Furthermore, the starting position of the X-ray tube, the geometry of the rotation and the image size and the detector size are know. Additionally, for the blob basis the blob parameters are known and the LUT of area integrals is pre-calculated. All in all, it is assumed that the following parameters are given or pre-calculated based on the acquisition geometry:

---
[1] The corresponding system matrix is transpose of the forward projection system matrix **A**.
[2] This is done because matrix indexing in MATLAB® starts at (1,1). In other languages (e.g. C++) it might be (0,0).

3.6 Efficient distance-driven projector in 2D

- $(x_t^{start}, y_t^{start})$ are coordinates of the X-ray tube in the starting position;
- $(x_d^{start}, y_d^{start})$ are vectors of the detector boundaries in the starting position;
- N_p is the number of pixels in a row (column) in the image;
- N_{det} is the number of detector elements;
- (x_{iso}, y_{iso}) are the coordinates of the rotation iso-center;
- θ is an angle between x-axis and the normal vector of the x-ray beam;
- lut_area_blob is the lookup table of the blob area integrals.

Before the distance-driven algorithm for the current view can start, the tube and the detector must be rotated around the iso-center by angle θ (e.g. anticlockwise)

$$\begin{cases} x = ((x^{start} - x_{iso})\cos\theta - (y^{start} - y_{iso})\sin\theta) + x_{iso} \\ y = ((x^{start} - x_{iso})\sin\theta - (y^{start} - y_{iso})\cos\theta) + y_{iso}. \end{cases} \quad (3.20)$$

In MATLAB® the vector of the detector boundaries can be efficiently rotated using the vectorized implementation.

Basically, the distance-driven routine for pixels consists of mapping the pixel and detector boundaries onto a common axis (Fig. 3.10a) resulting in a set of mapped pixel boundaries $\{p_j\}$ and a set of mapped detector boundaries $\{d_i\}$. The overlap length is used as a weighting for the contribution of the pixel $p_{j_1 j_2}$ into the detector entry $d_{i_1 i_2}$

$$\begin{cases} d_{12} = \frac{p_{12}(d_2 - p_1)}{(d_2 - d_1)} L \\ d_{23} = \frac{p_{12}(p_2 - d_2) + p_{23}(d_3 - p_2)}{(d_3 - d_2)} L \end{cases} \quad (3.21)$$

The algorithm for blobs is similar to the algorithm for pixels, however, some modifications have to be done. A LUT of line integrals must be pre-calculated and the overlapping nature of blobs has to be taken into account (Fig. 3.10b). Instead of the simple intersection length, a k-th subarea $S_k^{b_j}$ of the blob b_j is used as the weighting coefficient of the contribution of the blob b_j into a detector entry $d_{i_1 i_2}$

$$\begin{cases} d_{12} = \frac{S_1^{b_1}}{(d_2 - d_1)} \\ d_{23} = \frac{S_2^{b_1} b_1 + S_1^{b_2} b_2}{(d_3 - d_2)} \end{cases} \quad (3.22)$$

The pixel-based and blob-based distance-driven forward projector is summarized in algorithm 3.1 and algorithm 3.2.

Chapter 3. Forward and backprojections

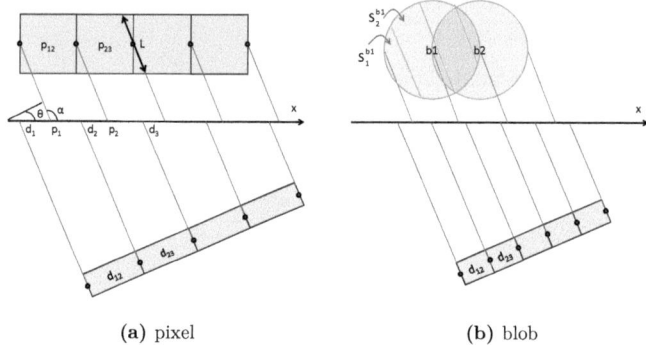

(a) pixel (b) blob

Figure 3.10: The distance-driven principle for (a) pixel basis function; (b) blob basis function.

Algorithm 3.1: Two dimensional distance-driven forward projector for pixel basis functions

Input: image, geometry description
Output: sinogram

1 **for** *all angular views θ* **do**
2   rotate the detector and the X-ray tube to the new position;
3   map detector boundaries onto the common axis;
4   **for** *all image rows* **do**
5   map pixel boundaries onto the axis;
6   **for** *for all projected points* **do**
7   find the length of each overlap ;
8   write the weighted pixel value into the detector element;

3.6.2 Unification of four angular cases

The choice of the common axis for boundaries mapping depends on the angular view θ

$$\begin{cases} \text{case 1}: & \frac{7\pi}{4} \leq \theta \leq \frac{\pi}{4} \quad \text{map to } x\text{-axis}; \\ \text{case 2}: & \frac{\pi}{4} < \theta < \frac{3\pi}{4} \quad \text{map to } y\text{-axis}; \\ \text{case 3}: & \frac{3\pi}{4} \leq \theta \leq \frac{5\pi}{4} \quad \text{map to } x\text{-axis}; \\ \text{case 4}: & \frac{5\pi}{4} < \theta < \frac{7\pi}{4} \quad \text{map to } y\text{-axis}. \end{cases} \quad (3.23)$$

Instead of implementation of four different code parts for each case, the algorithm can be generalized. In this subsection a trick will be shown how this generalization can

3.6 Efficient distance-driven projector in 2D

Algorithm 3.2: Two dimensional distance-driven forward projector for blob basis function

Input: image, geometry description, blob parameters, LUT
Output: sinogram

1. **for** *all angular views θ* **do**
2.   rotate the detector and the X-ray tube to the new position;
3.   map detector boundaries onto the common axis;
4.   **for** *all image rows* **do**
           // take into account the overlap of blobs ;
5.   map left boundaries onto the axis;
6.   map blob centers onto the axis;
7.   map right boundaries onto the axis;
8.   **for** *for all projected points* **do**
9.   find the length of each overlap and its location with respect to the blob center ;
10.   retrieve the x-ray integral from the LUT ;
11.   write the weighted pixel value into the detector element;

---

be implemented. Depending on the chosen axis, the calculations of all intersection points and the distance-driven routine require different formulas. However, it is possible to reduce the computation to a set of general formulas, which can be used in each case with properly adapted parameters. For the calculations two images are used: the original image `img` for the case of x-axis and the transposed and flipped image `img_y=flipud(fliplr(img'))` for the case of y-axis. Two mapping functions are defined `mapp2x` and `mapp2y`, see Listing 3.2. The mapping functions have to be used in the similar way to map the detector boundaries `detx=mapp(xt,yt,xd,yd)` and pixel boundaries `pix=mapp(xt,yt,xp,yp)`. Here, (`xt`, `yt`) is the X-ray tube position, (`xp`, `yp`) are vectors with the pixel boundaries coordinates and (`xd`, `yd`) are vectors with coordinates of the detector elements boundaries.

Listing 3.2: Matlab code for mapping function to x-axis and y-axis

```
1  function [y] = mapp2y(x1,y1,x2,y2)
2    y=y1-x1.*(y1-y2)./(x1-x2);
3  end
4  function [x] = mapp2x(x1,y1,x2,y2)
5    x=-(y1).*(x1-x2)./(y1-y2)+x1;
6  end
```

These functions construct a line connecting the X-ray source (`xt`, `yt`) and the detector (pixels) boundaries and extend this line until it intersects the x-axis (y-axis). The output of the functions (detx, pix) are vectors with those intersection points. This way, boundaries are mapped onto the axis.

The interception length L between the ray defined by θ and the image row is

$$\begin{cases} L = \frac{\Delta p}{\cos\theta}, & \text{if case 1 or case 3;} \\ L = \frac{\Delta p}{\sin\theta}, & \text{if case 2 or case 4.} \end{cases} \quad (3.24)$$

For the fan-beam geometry, beside the intersection length L of the central beam, the divergence has to be taken into account. For each beam in the fan a different intersection length must be calculated, see Listing 3.3.

Listing 3.3: Correction of the intersection length L for the divergent geometry

```
1  if case_x==true
2      for idx =1:ndet;
3          L1(idx)=sqrt((xt-detx(center_det)).^2+(yt-0).^2)...
4              /sqrt((xt-detx(idx)).^2+(yt-0).^2);
5      end
6  elseif case_y==true
7      for id =1:ndet;
8          L1(idx)=sqrt((xt-0).^2+(yt1-detx(center_det)).^2)...
9              /sqrt((xt-0).^2+(yt1-detx(idx)).^2);
10     end
11 end
12 L=L./L1;  % a correction is different for each beam in the fan
```

Beside the choice of the mapping axis and the length through the pixel L, the difference between those four cases is the order of the projected detector elements. In case 3 and case 4, the order of projected detector boundaries is reversed with respect to the order of projected pixel boundaries. The order of projected pixel boundaries always coincides with the positive direction of the common axis, while the order of the detector boundaries in the case 3 and the case 4 is opposite to the positive direction of the axis, see Fig. 3.11. Therefore, when running a loop over all intercepts[1], the starting detector index det_idx_start and the increment of the detector index det_idx_inc depend on the angular case

$$\begin{cases} \texttt{det\_idx\_start = 1,} & \texttt{det\_idx\_inc = +1,} & \text{if case 1 or case 2;} \\ \texttt{det\_idx\_start = N\_det,} & \texttt{det\_idx\_inc = -1,} & \text{if case 3 or case 4.} \end{cases} \quad (3.25)$$

The starting pixel index and the increment of the pixel index are always equal one

---

[1] The term *intercept* has been used by Bruno de Man in his original paper (De Man 2004)

3.6 Efficient distance-driven projector in 2D 63

`pix_idx_start = 1, pix_idx_inc = +1`.

It is defined that the index of the current detector corresponds to the index of its projected *left* boundary. However, in case 3 and case 4, the geometry is flipped and the starting boundary is the very *right* boundary. Therefore, in order to correct for this, two an additional parameters c_1 and c_2 are needed

$$\begin{cases} c_1 = 0,\ c_2 = 1 & \text{if case 1 or case 2} \\ c_1 = 1,\ c_2 = 0 & \text{if case 3 or case 4} \end{cases}. \quad (3.26)$$

Their role will be shown in the next subsection.

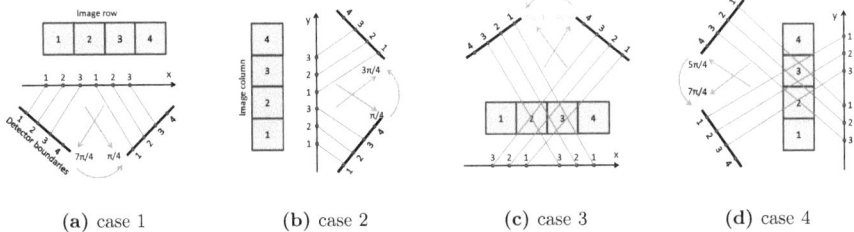

(a) case 1 (b) case 2 (c) case 3 (d) case 4

Figure 3.11: Drawings of four angular cases. The axis for mapping is the x-axis in the case 1 and case 3 and the y-axis in the case 2 and case 4. In case 3 and case 4, the order of projected detector boundaries is reversed with respect to the order of projected pixel boundaries and the positive direction of the axis.

3.6.3 The sweep line principle for pixels

The result of the mapping of the pixel boundaries and detector boundaries onto a common axis is two vectors of values, `rowx` and `detx`, correspondingly. To run a loop over the projected pixel boundaries and detector boundaries i.e. through all intersects, these two vectors must be merged together, i.e. via pre-sorting. In order to run the loop on-the-fly without pre-sorting, a concept of the *sweep line* or the *moving left boundary* is proposed. This line separates processed and non-processed mapped values. It always points to the left-hand side coordinate of the current overlapping interval. At each loop step, the running left boundary is either moved to the value of the pixel-related vector `rowx` or to the value of the detector-related vector `detx`. The decision where to move the line is done based on the relationships between the boundaries of the previous, current, and next pixel and the previous, current, and next detector. In the following the initialization of the sweep line and the moving algorithm will be described.

For the initialization of the sweep line, the first meaningful intercept must be detected. There are two cases possible: some detector boundaries mapped where no pixels are present (Fig. 3.12a) or some pixel boundaries are mapped outside the detector area (Fig. 3.12b).

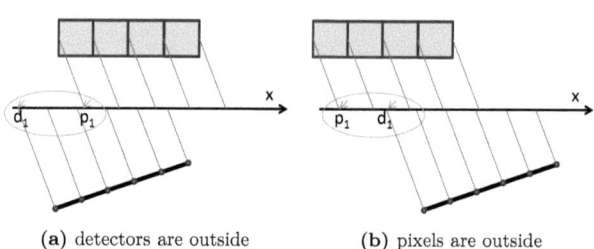

(a) detectors are outside (b) pixels are outside

Figure 3.12: Two cases after mapping the boundaries (a) some detector boundaries mapped where no pixels are present ; (b) some pixel boundaries are mapped outside detector area.

To identify these cases the distance between the first mapped detector boundary d_1=detx(det_idx_start) and the first mapped pixel boundary p_1=rowx(pix_idx) is calculated $d_1 - p_1$. If the distance is negative and its absolute value is larger than the size of a mapped detector element, then there is no overlap between any pixel and the current detector element. The sequential search through all mapped detector boundaries must be done until the first overlapping is detected and the $|d_1 - p_1|$ becomes smaller than the size of a mapped detector element, see Listing 3.4.

Listing 3.4: Moving left boundary is the left detector boundary
```
1  det_idx=det_idx_start; % start with ``first'' detector
2  if (detx(det_idx+c)-rowx(pix_idx))<-Δ_detx
3      while (detx(det_idx+c)-rowx(pix_idx))<-Δ_detx % no overlap
4          det_idx=det_idx+det_idx_inc; % go to next detector
5      end
6  end
```

Alternatively, if the distance $d_1 - p_1$ is positive and larger than one, it means that a sequential search through all pixel mapped boundaries is needed. The first overlap in this case will be when $d_1 - p_1$ becomes smaller than the size of projected pixel element, see Listing 3.5.

3.6 Efficient distance-driven projector in 2D

Listing 3.5: Moving left boundary is the left pixel boundary

```
1  pix_idx=1; %start with first pixel
2  if (detx(det_idx_start+c)-rowx(pix_idx))>Δ_pix
3      while (detx(det_idx_start+c)-rowx(pix_idx))>Δ_pix  % no overlap
4          pix_idx=pix_idx+1; % go to next pixel
5      end
6  end
```

After the index of the pixel and the detector in the first overlap is found, the value of `running_left_boundary` can be assigned, depending which mapped boundary (detector or pixel) has a smaller coordinate, see Listing 3.6.

Listing 3.6: Moving left boundary is the left pixel boundary

```
1  if detx(in_det_count+c1)≤rowx(pix_idx)
2      moving_left_boundary=rowx(pix_idx);
3  else
4      moving_left_boundary=detx(det_idx+c1);
5  end
```

After the start parameter of the `moving_left_boundary` is identified, the line can be moved through all intercept sequentially until all pixel boundaries and detector boundaries will be processed. While scanning though all mapped boundaries, either the detector index or the pixel index should be incremented. The sweep line always points to the left coordinate of the overlap. The next value of the sweep line is either the detector boundary or pixel boundary, depending which boundary is closer of the current position of the sweep line. Two cases are possible. In the first case, the still "unprocessed" part of the contribution of the current pixel is split between the current and next detector element(s), (Fig. 3.13a). The pixel index stays fixed, but the detector index in moved to the next detector element d'. The left boundary of d' is assigned to the value of the `moving_left_boundary`. In the second case, more than one pixel contribute to the current detector element, (Fig. 3.13b). The detector index stays fixed, but the pixel index is moved to the next pixel p'. The left boundary of p' is assigned to the value of the `moving_left_boundary`. The current position of the `moving_left_boundary` in Fig. 3.13 is depicted by a bold line and is in the position (1). The next position of the `moving_left_boundary` is the position (2). In order to account for the flipped detector, when θ is larger than $\frac{3\pi}{4}$, and still being able to have only one routine for the moving left boundary approach, the parameters $c1$ and $c2$, defined in the previous subsection, are used as it is shown in Listing 3.7.

Chapter 3. Forward and backprojections

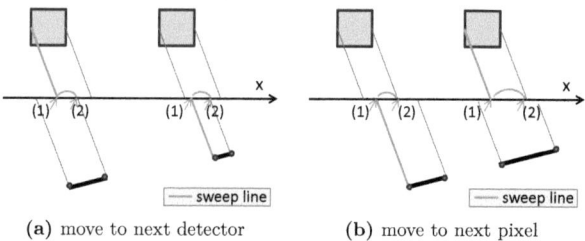

(a) move to next detector   (b) move to next pixel

Figure 3.13: While scanning though all mapped boundaries, two cases are possible which boundary will be the next moving_left_boundary. (a) it is moved to the next detector element; (b) it is moved to the next pixel

Listing 3.7: Moving left boundary approach to scan through the mapped boundaries

```
1  while ((pix_idx<Np)&&(det_idx<Ndet)&&(det_idx≥0))
2      if (detx(det_idx+c2)≤rowx(pix_idx+1))
3          % case A! we stay in the pixel, but move to next detector
4          o=detx(det_idx+c2)-left_boundary;  % overlap length
5          sino(det_idx)=sino(det_idx)+o*img(pix_idx);  % collect value
6          det_idx=det_idx+det_idx_inc;  % go to next detector
7          moving_left_boundary=detx(det_idx+c1);
8      else  % case B! we stay in the detector but move to next pixel
9          o=rowx(pix_idx+1)-left_boundary;  % overlap length
10         sino(det_idx)=sino(det_idx)+o*img(pix_idx);  % collect value
11         pix_idx=pix_idx+1;  % go to next pixel
12         moving_left_boundary=rowx(pix_idx);
13     end
14 end
```

In this routine

- `pix_idx` is an index of the current pixel (its left boundary);
- `det_idx` is an index of the current detector element (its left boundary);
- `N_p, N_det` are total number of pixels and detector elements;
- `N_p+1, N_det+1` is the total number of pixel and detector boundaries, respectively;
- `rowx, detx` is the mapped pixel row and detector boundaries, respectively;
- `moving_left_boundary` is the sweep line;
- `o` is the calculated overlap length;
- `c1, c2` are parameters which accounts for all four angular cases;

3.6.4 The sweep line principle for blobs

The sweep line principle for pixels can be used with minor modifications for blob basis functions as well. In contrast to the pixel case, where the right boundary of pixel p_i coincides with the left boundary of the next pixel p_{i+1}, the left and right boundaries of the blob do not have such symmetry and have to be addressed separately. The mapping of the blobs results in three vectors `blobx_left`, `blobx_ritgh` and `blobx_cntr`, which are the mapped left boundaries, the mapped right boundaries and the mapped blob centers, accordingly. The mapped blob centers are needed to retrieve the area projection integral from the pre-calculated LUT. This can be done, e.g. using the nearest neighbor interpolation.

To find the first value of the `moving_left_boundary`, the vector of left blob boundaries must be compared with the vector of detector boundaries similar to Listing 3.4 and Listing 3.5. Then, the same as in the algorithm for pixel basis functions, two cases are possible: the detector index is incremented or the pixel index is incremented. The algorithmic part for moving to the next detector is identical to the pixel case. The algorithm for moving to the next blob must be modified taking into account the overlapping nature of blobs. From the detector point of view, moving to the next blob means moving backwards in the detector index, because the left boundary of the next blob is on the left-hand side with respect to the right boundary of the blob, which is just left. A MATLAB® code for the modified moving left boundary approach is given in Listing 3.8.

Listing 3.8: Modified moving left boundary approach for blobs

```
1  det_sub_idx=det_idx; % start with first detector
2  while (pix_idx<Np)&&(det_sub_idx≤Ndet)&&(det_idx≤Ndet)&&...
3         (det_sub_idx≥1)&&(det_idx≥1)
4      % scaling factor from projected blob size to original blob size
5      shrink=(2*basis_param.blob_radius/shift_pix_test(pix_idx));
6      if detx(det_sub_idx+c2)≤blobx_right(pix_idx)
7          % case A! we stay in the blob, but move to the next detector
8          % right-side  ditance to the blob center
9          right_idx=(detx(det_sub_idx+c2)−blobx_cntr(pix_idx))*shrink;
10         % left-side ditance to the blob center
11         left_idx=(running_left_boundary−blobx_cntr(pix_idx))*shrink;
12         % transform distances to index from LUT, e.g. nearest neighbor
13         right_idx_from_lut=fix((right_idx+blob_radius)/∆_lut)+1;
14         left_idx_from_lut=fix((left_idx+blob_radius)/∆_lut)+1;
15         % retrieve values from LUT
16         left_area=lut_area_blob(up_idx_from_lut);
17         right_area=lut_area_blob(low_idx_from_lut);
18         o=right_area−left_area; % strip area
19         sino(det_idx)=sino(det_idx)+o*img(pix_idx); % collect value;
20         det_sub_idx=det_sub_idx+det_idc_inc; % go to next detector
21         running_left_boundary=detx(det_sub_idx+c);
22     else % case B! we stay in the detector but hmove to next blob
23         % Calulate the overlapping area 'o' as it is shown in case A
24         right_idx=(detx(det_sub_idx+c2)−blobx_cntr(pix_idx))*shrink;
25         left_idx=(running_left_boundary−blobx_cntr(pix_idx))*shrink;
26         right_idx_from_lut=fix((right_idx+blob_radius)/∆_lut)+1;
27         left_idx_from_lut=fix((left_idx+blob_radius)/∆_lut)+1;
28         left_area=lut_area_blob(up_idx_from_lut);
29         right_area=lut_area_blob(low_idx_from_lut);
30         o=right_area−left_area;
31         sino(det_idx)=sino(det_idx)+o*img(pix_idx); % collect value;
32         pix_idx=pix_idx+1; % go to next pixel
33         running_left_boundary=blobx_left(pix_idx+s);
34         while detx(det_idx+c2)≤rowx_left(pix_idx)
35             % go backwards to the "first detector" for current blob
36             det_idx=det_idx+det_idc_inc;
37         end
38         det_sub_idx=det_idx; % go back to "1st" sub-detector
39     end
40 end
```

3.7 Efficient distance-driven projector in 3D

3.7.1 General strategy

The distance-driven forward projector and backprojector can be efficiently extended to the third dimension. In the first step of the three-dimensional distance-driven projector, all voxel and detector element boundaries are mapped onto a common plane, e.g. xz-plane according to the imaging geometry, see Fig. 3.14a. In practice, pixels and detector elements are mapped through their horizontal and vertical boundaries approximating their shape as a rectangle. Pixel and detector boundaries in x-direction and z-direction are mapped separately. Pixels are processed plane-by-plane. The area of an overlap is given by the multiplication of the overlap in x- and z-axis. The area of the overlap is used as the weighting coefficient for the contribution of the j-th voxel to the i-th detector in forward- and backprojectors, see algorithm 3.3. The assumptions from the two-dimensional case regarding the acquisition geometry and known parameters are valid. The coordinates of the starting position of the X-ray tube are known and the coordinates of the detector boundaries are pre-computed based on the detector type.

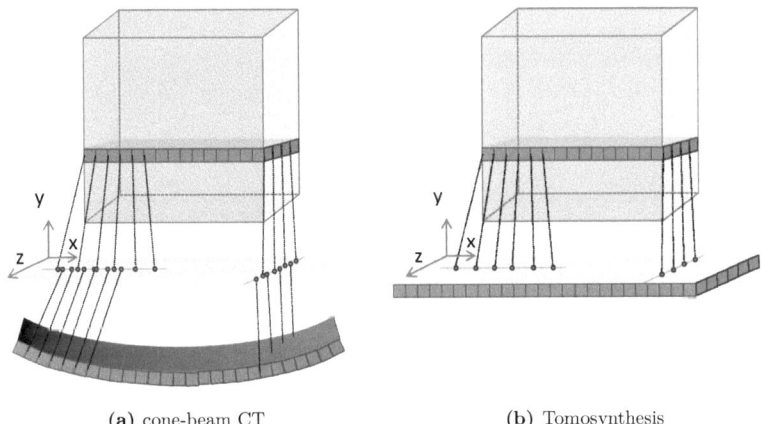

(a) cone-beam CT   (b) Tomosynthesis

Figure 3.14: Distance-driven projector in three dimensions. (a) in cone-beam CT pixel and detector boundaries are mapped onto a common plane; (b) in tomosynthesis with fixed-detector geometry, the detector plane can be used as a common plane and only pixel boundaries must be mapped.

Moreover, the distance-driven approach can benefit from the tomosynthesis geometry with a fixed detector. Here, the detector plane can be used as the common plane for the mapping. In Fig. 3.14b one can see that the detector lies in the xz-plane. The mapping of the detector element boundaries is not needed anymore, which makes the projector computationally more efficient, see algorithm 3.4. Mapping of the pixels may be considered as an isotropic stretch operator, while no change in the shape occurs i.e. the mapped pixels remain square. Therefore, the mapping operation can be replaced by the less expensive incremental operation.

Algorithm 3.3: Three-dimensional distance-driven forward projector for pixel basis functions and cone-beam CT geometry.

Input: image, geometry description
Output: sinogram

1 **for** *all angular views θ* **do**
2   rotate the detector and the X-ray tube to the new position;
3   map detector x-boundaries onto the common plane;
4   map detector z-boundaries onto the common plane;
5   **for** *all image y-planes* **do**
6   map pixel x-boundaries onto the common plane;
7   map pixel z-boundaries onto the common plane;
8   **for** *for all projected points* **do**
9   find the length of x-overlap;
10   find the length of z-overlap;
11   find the area of overlap;
12   write the weighted pixel value into the detector element;

Algorithm 3.4: Three-dimensional distance-driven forward projector for pixel basis functions and digital tomosynthesis geometry with fixed flat-panel detector.

Input: image, geometry description
Output: sinogram

1 **for** *all angular views θ* **do**
2   rotate the the X-ray tube to the new position;
3   **for** *all image y-planes* **do**
4   map pixel x-boundaries onto the detector plane;
5   map pixel z-boundaries onto the detector plane;
6   **for** *for all projected points* **do**
7   find the length of x-overlap;
8   find the length of z-overlap;
9   find the area of overlap;
10   write the weighted pixel value into the detector element;

3.7 Efficient distance-driven projector in 3D

3.7.2 Unification of the angular cases

The choice of the common plane in general case depends on the angle between the middle ray in the cone and y-axis (similarly to 2D case)

$$\begin{cases} \text{case 1}: & \frac{7\pi}{4} \leq \theta \leq \frac{\pi}{4} \quad \text{map to } xz\text{-plane}; \\ \text{case 2}: & \frac{\pi}{4} < \theta < \frac{3\pi}{4} \quad \text{map to } yz\text{-plane}; \\ \text{case 3}: & \frac{3\pi}{4} \leq \theta \leq \frac{5\pi}{4} \quad \text{map to } xz\text{-plane}; \\ \text{case 4}: & \frac{5\pi}{4} < \theta < \frac{7\pi}{4} \quad \text{map to } yz\text{-plane}. \end{cases} \qquad (3.27)$$

The mapping functions can be unified (Listing 3.9) and used e.g. to map detector boundaries as `[detx,detz] = mapp(xt,yt,zt,xd,yd,zd)`. For the further explanations, all variables related to the projected detector boundaries, e.g. `detx` and `dety` will be noted by the ending x, meaning it is accounted for projection to both, xz- and yz-planes by adjusting the parameters in the preamble of the algorithm.

Listing 3.9: Vectorized MATLAB® code for the mapping function to the xz-plane and the yz-plane in three dimensions

```
1  function [x,z] = mapp2xz(x1,y1,z1,x2,y2,z2)
2    x=(-y1).*(x2-x1)./(y2-y1)+x1;
3    z=(-y1).*(z2-z1)./(y2-y1)+z1;
4  end
5  function [y,z] = mapp2yz(x1,y1,z1,x2,y2,z2)
6    y=(-x1).*(y2-y1)./(x2-x1)+y1;
7    z=(-x1).*(z2-z1)./(x2-x1)+z1;
8  end
```

To account for the "flip" of detector boundaries in case 3 and case 4, the value of the starting detector index `det_idx_x_start` and the increment of this index `det_idx_x_inc` must be chosen based on the angular view

$$\begin{cases} \texttt{det\_idx\_x\_start} = 1, & \texttt{det\_idx\_x\_inc} = +1, \quad \text{if case 1 or case 2}; \\ \texttt{det\_idx\_x\_start} = \texttt{N\_detx}, & \texttt{det\_idx\_x\_inc} = -1, \quad \text{if case 3 or case 4}. \end{cases} \qquad (3.28)$$

There is no detector boundary flip in z-direction, therefore the starting detector index `det_idx_z_start` and its increment `det_idx_z_inc` are always equal one.

Additionally, the same as in the two-dimensional case, several additional parameters are needed to account for the reverse of the boundaries order in x-direction

$$\begin{cases} cx_1 = 0,\ cx_2 = 1,\ cz_1 = 0,\ cz_2 = 1 & \text{if case 1 or case 2}; \\ cx_1 = 1,\ cx_2 = 0,\ cz_1 = 0,\ cz_2 = 1 & \text{if case 3 or case 4}. \end{cases} \qquad (3.29)$$

Boundary order in z-direction stays always the same, but the z-related parameters cz_1 and cz_2 are also introduced to make the algorithm symmetrical.

For tomosynthesis limited angle geometry the parameters from the case 1 are applicable.

3.7.3 Sweep line in three dimensions

The mapping of pixel boundaries and detector boundaries in the general three-dimensional case results in four vectors: `detx`, `rowx`, `detz` and `rowz`. To run a loop over all intercepts, we extend the concept of the sweep line is extended to the third dimension and two running boundaries are used simultaneously.

The variables `left_boundaryX`, `left_boundaryZ` describes the sweep line in x-direction and in z-direction, correspondingly. The loop over the x-intercepts is nested into the loop over z-intercepts. First of all, the first value for two running boundaries must be identified. The algorithms from Listing 3.5 and Listing 3.4 for the two-dimensional case can be directly applied for both, x- and z-directions. The algorithm for the moving boundary to scan through all mapped boundaries can also be directly taken from the two-dimensional case, see Listing 3.7. It results in the following routine for a distance-driven projector in three dimensions, see Listing 3.10.

3.7 Efficient distance-driven projector in 3D

Listing 3.10: The distance-driven routine for three dimensional cone-beam CT. The sub-routine for the z-axis is in Listing 3.11

```
1   % Initialization of x-axis―――――――――――――――≫
2   pix_idx_x=1;det_idx_x=det_idx_x_start;
3   if (detx(det_idx_x)-rowx(pix_idx_x))>pix_size_x
4       while ((detx(det_idx_x)-rowx(pix_idx_x))>pix_size_x)
5           pix_idx_x=pix_idx_x+1;
6       end
7   end
8   if (detx(det_idx_x)-rowx(pix_idx_x))<-det_size_x
9       while (detx(det_idx_x)-rowx(pix_idx_x))<-det_size_x
10          det_idx_x=det_idx_x+det_idx_x_inc;
11      end
12  end
13  if detx(det_idx_x)<rowx(pix_idx_x)
14      left_boundaryX=rowx(pix_idx_x);
15  else
16      left_boundaryX=detx(det_idx_x+sx);
17  end
18  %Loop over x-axis―――――――――――――――――≫
19  while ((pix_idx_x<Nx)&&(det_idx_x<ndetz)&&(det_idx_x≥det_idx_x_start))
20      if (detx(det_idx_x+c2x)<rowx(pix_idx_x+1))    %A!
21          oX=(detx(det_idx_x+c2x)-left_boundaryX)/det_size_x;
22      else %B!
23          oX=(rowx(pix_idx_x+1)-left_boundaryX)/det_size_x;
24      end
25
26      % <The sub-routine for z-axis is here>
27
28      if (detx(det_idx_x+c2x)≤rowx(pix_idx_x+1))   %A!
29          det_idx_x=det_idx_x+det_idx_x_inc;
30          left_boundaryX=detx(det_idx_x);
31      else %B!
32          pix_idx_x=pix_idx_x+1;
33          left_boundaryX=rowx(pix_idx_x);
34      end
35  end
```

Listing 3.11: MATLAB® sub-routine for the z-axis. It is a part of the Listing 3.10.

```
1   % Initialization of z-axis―――――――――――――>>
2   pix_idx_z=0;det_idx_z=det_idx_z_start;
3   if (((detz(det_idx_z)-rowz(pix_idx_z)))>pix_size_z)
4       while (((detz(det_idx_z)-rowz(pix_idx_z)))>pix_size_z)
5           pix_idx_z=pix_idx_z+1;% go to next pixel column
6       end
7   end
8   if (detz(det_idx_z)-rowz(pix_idx_z))<-det_size_z
9       while (detz(det_idx_z)-rowz(pix_idx_z))<-det_size_z
10          det_idx_z=det_idx_z+go_to_next_detz;% go to next z-detector
11      end
12  end
13  if detz(det_idx_z+sz)<rowz(pix_idx_z)
14      left_boundaryZ=rowz(pix_idx_z);
15  else
16      left_boundaryZ=detz(det_idx_z);
17  end
18  %Loop over z-axis―――――――――――――>>
19  while (pix_idx_z<nz)&&(det_idx_z<ndetz)&&(det_idx_z≥det_idx_z_start)
20      if (detz(det_idx_z+c2z)≤rowz(pix_idx_z+1))% A!
21          oZ=(detz(det_idx_z+c2z)-left_boundaryZ)/det_size_z;
22          sino(det_idx_x*ndetz+det_idx_z)=...
23              sino(det_idx_x*ndetz+det_idx_z)+...
24              oZ*oX*slice_y(pix_idx_x*nz+pix_idx_z);
25          det_idx_z=det_idx_z+go_to_next_detz;%% go to the next detector
26          left_boundaryZ=detz(det_idx_z);% save where did we come from
27      else    %% B!
28          oZ=(rowz(pix_idx_z+1)-left_boundaryZ)/det_size_z;
29          sino(det_idx_x*ndetz+det_idx_z)=...
30              sino(det_idx_x*ndetz+det_idx_z)+...
31              oZ*oX*slice_y(pix_idx_x*n3+pix_idx_z);
32          pix_idx_z=pix_idx_z+1;%% go to the next pixel
33          left_boundaryZ=rowz(pix_idx_z);% save where did we come from
34      end
35  end
```

Chapter 4

Iterative image reconstruction for tomosynthesis

Contents

4.1	Discrete model of the physical system	76
4.2	Iterative reconstruction schemes	77
4.3	Considerations for practical implementation of SART . . .	82
4.4	Projection access order for SART	87

The task of the tomographic reconstruction algorithm is to recover the unknown distribution of the X-ray attenuation coefficients based on the measured data. In this chapter a family of iterative reconstruction algorithms will be discussed. First, the formulation of the discrete model will be introduced and the reconstruction problem will be formulated as an optimization problem. Second, two types of iterative reconstruction algorithms, namely algebraic and statistical reconstruction will be presented.

The main focus of this chapter will be put on the simultaneous algebraic reconstruction technique (SART) with application to tomosynthesis data. Considerations for the practical implementation, will be given, including the data handling strategy and memory and computation costs. Finally, the projection access order for SART will be discussed. A novel correlation-based scheme will be presented. It will be compared with existing schemes using a simulation study.

4.1 Discrete model of the physical system

As it has been mentioned in Chapter 2, the measurement p through a continuous object $\mu(\mathbf{x})$ along a line L defined by the direction θ can be written as a line integral

$$p(\theta) = \int_{L(\theta)} f(\mathbf{x}) \, d\mathbf{x}. \tag{4.1}$$

The continuous object has to be discretized. This can be done using a set of a finite number of coefficients c_i and basis functions $b_i(\mathbf{x} - \mathbf{x_i})$, as it has been discussed in Chapter 3

$$f(\mathbf{x}) \approx \sum_{i=1}^{N} c_i b_i (\mathbf{x} - \mathbf{x_i}). \tag{4.2}$$

Combining equation 4.1 and equation 4.2 leads to

$$p_j = \int_{L(\theta_j)} \sum_{i=1}^{N} c_i b_i (\mathbf{x} - \mathbf{x_i}) \, d\mathbf{x}, \tag{4.3}$$

where j is an index of the measured beam. The order of the integration and the summation can be changed and the coefficients c_i can be taken out of the integral as constants

$$p_j = \sum_{i=1}^{N} c_i \int_{L(\theta_j)} b_i (\mathbf{x} - \mathbf{x_i}) \, d\mathbf{x}. \tag{4.4}$$

The line integral through the basis functions b_i can be pre-computed or calculated on-the-fly. This results in the discrete model of the measurements

$$p_j = \sum_{i=1}^{N} c_i a_{ij} \text{ with } a_{ij} = \int_{L(\theta_j)} b_i (\mathbf{x} - \mathbf{x_i}) \, d\mathbf{x}. \tag{4.5}$$

If this is done for all measured beams, the complete measurement process can be written as a system of linear equations

$$\mathbf{p} = \mathbf{Af}. \tag{4.6}$$

Here, $\mathbf{f} = (f_1, ..., f_N)^{\mathbf{T}} \in \mathbb{R}^N$ is a discrete representation of the three-dimensional spatial distribution of the X-ray linear attenuation coefficients within the imaged volume, $\mathbf{p} = (p_1, ..., p_M)^{\mathbf{T}} \in \mathbb{R}^M$ is a vector of the measured projection data and $\mathbf{A} \in \mathbb{R}^{M \times N}$ is a system matrix. N is the total number of image volume elements (voxels) and M is the total number of measured beams. M is equal to the number of detector elements (dexels) N_d multiplied by the number of views N_θ. The system matrix $\mathbf{A}$ describes the measurement process and the acquisition geometry and represents line integrals though the basis functions. It is a flexible way to describe any arbitrary acquisition geometry setup. A system matrix element a_{ij} represents the contribution of the i-th voxel f_i to the j-th beam p_j, i.e. it models the discrete forward operator, see Chapter 3. The transpose of system matrix $\mathbf{A^T}$ describes the discrete backprojection operator. In practice, the system matrix is very large and sparse.

4.2 Iterative reconstruction schemes

It is important to note that although the line integrals $\mathbf{p}$ are measured, the image $\mathbf{f}$ is of interest

$$\mathbf{f} = \mathbf{A}^{-1}\mathbf{p}. \tag{4.7}$$

However, the direct matrix inversion of this equations is practically infeasible (Kak 1984) due to the large size of the equation system. Moreover, in the case of limited angle tomosynthesis acquisition geometry, this system is severely under-determined. This leads to the need of an alternative algorithm. One such approach is to formulate an optimization problem, which minimizes some pre-defined cost function Ψ to find the "best" solution $\hat{\mathbf{x}}$

$$\hat{\mathbf{x}} = \underset{\mathbf{x} \geq 0}{\operatorname{argmin}} \Psi(\mathbf{x}). \tag{4.8}$$

The cost function is constructed of two components: a *data-mismatch* term and a *regularization* or a *penalty* term with a regularization parameter β

$$\Psi(\mathbf{x}) = \text{DataMismatch}(\mathbf{p}, \mathbf{Ax}) + \beta \, \text{Regularization}(\mathbf{x}). \tag{4.9}$$

One could solve the minimization problem analytically by zeroing the gradient of $\Psi(x)$. Unfortunately, even if the non-negativity constraint and the penalty term are omitted, there are usually no closed-form solutions (Fessler 2000). Thus, the minimization problem must be solved iteratively. Under the ideal conditions the cost function completely defines the resulting image and independent of the choice of a minimization algorithm (De Man 2005). Two common minimization approaches are the conjugate gradient (CG) method (Mumcuoglu 1994, Piccolomini 1999, Fessler 1999) and the iterative coordinate descent (ICG) (Sauer 1993). The CG updates all pixels simultaneously, while the ICG updates one pixel at a time. It is expected that the global minimum of the cost function should be found regardless of the minimization approach, however, some algorithms could get stuck in a limit cycle behavior (local minimum) (De Man 2005).

4.2.1 Algebraic reconstruction

The algebraic reconstruction technique (ART) was first introduced to the field of image reconstruction by Gordon et. al. in 1970 (Gordon 1970)[1] for the application of the three-dimensional electron microscopy. ART is mathematically similar to the Kaczmarz algorithm for solving a system of linear equations (Kaczmarz 1937). An alternative method called simultaneous iterative reconstruction technique (SIRT) was proposed by P. Gilbert in 1972 (Gilbert 1972). Later, A. H. Anderson and A. C. Kak (Andersen 1984) proposed a superior version of ART and called it simultaneous algebraic reconstruction technique. Later, the authors applied SART to limited view CT (Andersen 1989). The ART algorithm typically results in noisy images while the SIRT has good noise properties but slow convergence. The SART algorithm results in the reduction of the noise associated with ART-type methods, while it preserves the convergence speed of ART-type methods.

All algebraic algorithms iteratively minimize the residual error between the measured data and the calculated forward projection of the estimated image.

A single update step of ART is given by

$$\mathbf{f}^{(n+1)} = \mathbf{f}^{(n)} + \frac{\mathbf{a_j}^T}{\mathbf{a_j}^T \mathbf{a_j}} \left(p_j - \hat{p}_j \left(\mathbf{f}^{(n)} \right) \right)$$
$$\hat{\mathbf{p}}(\mathbf{f}) = \mathbf{A}\mathbf{f}$$
(4.10)

and one update of SART is given by

$$f_i^{(n+1)} = f_i^{(n)} + \frac{1}{A_{i,+}} \sum_{j \in J_\theta} \frac{a_{i,j}}{A_{+,j}} \left(p_j - \hat{p}_j \left(\mathbf{f}^{(n)} \right) \right)$$
$$A_{i,+} = \sum_{j \in J_\theta} a_{i,j}$$
$$A_{+,j} = \sum_{i=1}^{N} a_{i,j}$$
$$\hat{\mathbf{p}}(\mathbf{f}) = \mathbf{A}\mathbf{f}.$$
(4.11)

Here, $\overline{\mathbf{p}}(\mathbf{f})$ is the forward projection of the current estimated image, $\mathbf{a_j}$ is a column of the matrix A, which describes j-th measured beam, $A_{i,+}$ is a normalization for numbers of rays intersecting each voxel, $A_{+,j}$ is a normalization for the path through all pixels in the current ray. In the n-th iteration a forward projection of the current estimated image is calculated. Then, it is compared with the measured data. Based on this, an updating term is calculated. Afterwards, the updating term is homogeneously backprojected into the image domain according to the system matrix **A**. The ART updating term

---

[1] However, it is also believed that G. Hounsfield also used ART in his scanner (Mueller 1998a)

4.2 Iterative reconstruction schemes

is applied for all pixels which contribute to the j-th ray, the SART updating term is applied for all pixels in the image considering one projection view $j \in J_\theta$.

It can be shown, that SART minimizes a weighted least squares (WLS) functional, but without taking into account any data noise models (Jiang 2003a)

$$L(f) = \sum_{j=1}^{M} \frac{1}{A_{+,j}} \left(p_j - \hat{p}_j\left(\mathbf{f}^{(\mathbf{n})}\right)\right)^2 = \|\mathbf{p} - \mathbf{Af}\|_w^2 \qquad (4.12)$$

The weighting term $w = \frac{1}{A_{+,j}}$ is in terms of system matrix elements and does not take any statistics into account. Papers which address the convergence of algebraic reconstruction technique include (Jiang 2003b, Wang 2007, Qu 2009). Further papers which address algebraic reconstruction methods include (Kak 1988, Toft 1996, Mueller 1998a, van de Sompel 2007, Nikazad 2008).

4.2.2 Statistical Reconstruction

Statistical reconstruction algorithms are based on statistical models and are equivalent to an optimization problem

$$\text{reconstructed image} = \underset{\text{image}}{\operatorname{argmax}}[\,P(\text{image}|\text{measurements}) + P(\text{image})]. \qquad (4.13)$$

The first term is a likelihood term. It defines the probability to obtain the measurements given the image. It is supposed, that the actual measured number of photons y_j include the noise and deviate from their expected values $\hat{y}_j$. The image is defined by the physical model of the measurement means

$$\hat{y}_j = I_0 e^{-\hat{p}_j}, \text{ where } \hat{p}_j = \sum_j a_{ij} f_i. \qquad (4.14)$$

The joint probability of all measurements is given by

$$P(\text{image}|\text{measurements}) = \prod_j P(y_j|\hat{y}_j) \qquad (4.15)$$

The log-likelihood is given by

$$\log P(\text{image}|\text{measurements}) = \log \prod_j P(y_j|\hat{y}_j) = \sum \log P(y_j|\hat{y}_j). \qquad (4.16)$$

The second term in equation 4.13 is called the prior term. The prior term controls the smoothness of the reconstructed image. Without the prior information, this optimization problem is called the *maximum likelihood* (ML) approach. With the prior term, this optimization problem is referred to as the *maximum a posteriori* (MAP) approach.

The task is to find the image, which maximizes the probability to obtain the given set of measured data. The likelihood term of the cost function is derived based on the noise model. Two common choices for the noise models are the Gaussian model and the Poisson model.

For the logarithmically processed data a Gaussian noise model is used. The Gaussian noise model is given by

$$P(p_j|\hat{p}_j)_{\text{Gaussian}} = \frac{1}{\sqrt{2\pi}\sigma_j} \exp\left(-\frac{(p_j - \hat{p}_j)^2}{2\sigma_j^2}\right), \qquad (4.17)$$

where σ is the standard deviation. Suppose, the value $\hat{p}_j$ is the expected projection value, which is calculated based on the physical model of measurements (Beer-Lambert-Law). However, because of noise a value p_j is measured. The probability to measure any particular value p_j is described by the Gaussian distribution. Then, the log-likelihood term for Gaussian noise model is given by

$$\log L(f, p) = \sum_{j=1}^{M} \log(P(p_j|\hat{p}_j)_{\text{Gaussian}}), \qquad (4.18)$$

where

$$\log(P(p_j|\hat{p}_j)_{\text{Gaussian}}) = \log(\sqrt{2\pi}\sigma_j) - \frac{1}{2}\frac{(p_j - \hat{p}_j)^2}{\sigma_j^2}. \qquad (4.19)$$

The constant term can be ignored. This leads to the weighted least-squares cost function which has to be minimized

$$\Psi_{\text{WLS}} = \frac{1}{2}\sum_{j=1}^{M} w_j (p_j - \hat{p}_j)^2 \text{ with } w_j = 1\big/\sigma_j. \qquad (4.20)$$

The weights w_j are represented by the reciprocal of the variance $1/\sigma_j^2$ of the log-data. However, in contrast to algebraic WLS, the weighting here takes into account the data statistics.

Another choice for the noise model is the Poisson noise model

$$P(y_j|\hat{y}_j)_{\text{Poisson}} = \frac{e^{-\hat{y}_j}\hat{y}_j^{y_j}}{y_j!}, \qquad (4.21)$$

where $\hat{y}_j$ is the expected number of photons and y_j is the actual number of measured

4.2 Iterative reconstruction schemes

photons. The data mismatch term, or the the log-likelihood in this case is

$$\log(L(y_j|\hat{y}_j)) = \sum_{j=1}^{M} h_j([\mathbf{Af}]_j), \qquad (4.22)$$

where

$$h_j = \log(P(y_j|\hat{y}_j)_{\text{Poisson}}), \qquad (4.23)$$

which results in

$$\begin{aligned} h_j &= \log(e^{-\hat{y}_j}\hat{y}_j^{y_j}) - \log(y_j!) \\ &= -\hat{y}_j + y_j \log(\hat{y}_j) - \log(y_j!)). \end{aligned} \qquad (4.24)$$

According to J. Fessler (Fessler 2000) the *expectation maximization* (EM) algorithm derived for emission tomography (Shepp 1982) is not very suitable for the Poisson transmission reconstruction problem (Lange 1984). Transmission EM-algorithms have slow convergence rate and high computation costs because of the large of number of exponential terms.

As an example, the formula of the ordered subset ML algorithm for transmission tomography (OS-MLTR) proposed by J. Nuyts (Nuyts 1997) is given by

$$f_i^{(n+1)} = f_i^{(n)} + \frac{\sum\limits_{j \in \text{subset}} a_{ij}(\hat{y}_j - y_j)}{\sum\limits_{j \in \text{subset}} a_{ij}\hat{y}_j \sum\limits_{k} a_{kj}}. \qquad (4.25)$$

The formula of ordered subset separable paraboloidal surrogates (OS-SPS) algorithm proposed by J. Fessler (Fessler 1997b, Erdogan 1999) is given by

$$f_i^{(n+1)} = f_i^{(n)} + \frac{\sum\limits_{j \in \text{subset}} a_{ij}(\hat{y}_j - y_j)}{\sum\limits_{j \in \text{subset}} a_{ij}y_j \sum\limits_{k} a_{kj}}. \qquad (4.26)$$

For a detailed discussion of the EM-algorithms for transmission tomography as well as more efficient algorithms for direct likelihood maximization (Nuyts 1997) such as coordinate-ascent (Fessler 1997b) and paraboloidal surrogates algorithms see the chapter 1 from the Handbook of medical imaging written by J. Fessler (Fessler 2000).

4.3 Considerations for practical implementation of SART

In the previous section algebraic and statistical reconstruction algorithms have been introduced. As the work from Zhand et al. (Zhang 2006b) shows, in practice, the differences in image quality between two types of algorithms are not large. However the statistical reconstruction algorithms usually need ten to twenty iterations, while the algebraic reconstruction provides acceptable image quality already after a few iterations. It was decided to use the SART algorithm in this thesis for reasons of simplicity and especially of the convergence speed. This section presents several aspects of the practical implementation of the SART algorithm in MATLAB® for three-dimensional tomosynthesis data. First, a dictionary approach and a base workspace will be discussed. Second, the memory costs and the computation complexity of SART will be approximated.

4.3.1 How to address tomosynthesis datasets: a dictionary approach

A tomosynthesis acquisition results in N_θ two-dimensional projections and a tomosynthesis reconstruction results in N_{slices} two-dimensional images. A procedure which can address automatically any of the projection images and reconstructed images is required. Any set of two-dimensional images can be stacked in MATLAB® into a volume resulting in a three-dimensional variable. Then, each slice from this set can be easily addressed using the corresponding index. Taking into account the size of tomosynthesis data (several GB) and typical available RAM memory of a PC (4 – 16 GB), stacking all images into a volume becomes infeasible due to memory limitations. Storing each projection and each reconstructed image as a separate variable provides more flexibility in terms of memory because one part of the dataset (e.g. a single slice) can be loaded, processed at a time and then saved on the disc and deleted from RAM afterwards to free memory for the next portion of data. Addressing large number of variables automatically can be done using a *dictionary approach*. A dictionary contains a list of all variable names. It is created once and then can be used to address any variable. To each projection and reconstruction dataset a name is assigned. The number of each projection (slice) image is not an index of volume anymore but it is stored as a part of the name of variable. Each projection (image) name in the dictionary consist of the dataset name and the corresponding number. An example of the dictionary function for reconstructed images is presented in Listing 4.1.

4.3 Considerations for practical implementation of SART

Listing 4.1: MATLAB® function for creating a dictionary

```
1  function dict=make_dictionary_for_slices(slice_name, nx, ny, nz_max)
2  dict=cell(nz_max,1);
3  for nz=1:nz_max
4      if nz≥1&&nz≤9
5          dict{nz}=sprintf('%s_%g_%g_slice00%g', slice_name, nx, ny, nz);
6      elseif nz≥10&&nz≤99
7          dict{nz}=sprintf('%s_%g_%g_slice0%g', slice_name,nx, ny, nz);
8      elseif ny≥100&&nz≤999
9          dict{nz}=sprintf('%s_%g_%g_slice%g', slice_name, nx, ny, nz);
10     end
11 end
```

A typical task is to access an i-th image, to perform some calculations and to save the result into the same variable. The name of the variable can be obtained from the dictionary based on the dataset name and the the value of the index i. Reading a variable, which name is saved as a string, can be done using the MATLAB® function `eval`. To perform the calculations, the obtained value of the variable should be saved in a temporary variable. After calculations are done, the value of the temporary variable has to be written back into the slice variable. This also can be done using the function `eval`, see Listing 4.2.

Listing 4.2: MATLAB® code for using a dictionary and function `eval`

```
1  slice_name = dict_reco_name{i}; % get the name from dictionary
2  f = eval(slice_name);           % write slice into a temporary variable f
3  f = ...% do some calculatins
4  eval([slice_name ' = f']) % save result f into the slice variable name
```

4.3.2 Memory handling in MATLAB®: the base workspace

All variables which have been created during a MATLAB® session are stored in a *workspace*. The variables, which have been created using a command line or the scripts are stored in the *base* workspace. When an arbitrary function `func` is called in MATLAB®, then all passed to the function variables are duplicated in a temporary workspace corresponding to this function. After all calculations are done and function returns its output, the temporary workspace is deleted. When processing tomosynthesis data, it is desired to avoid creating unnecessary copies of data in order to use most of the memory efficient. This can be done using the functions `assignin` and `evalin` which are used to access variables from the *base* workspace when a function `func` is executing, see Listing 4.3.

Chapter 4. Iterative image reconstruction

Listing 4.3: MATLAB® code for using the functions evalin and assignin

```
1  slice_name = dict_reco_name{i};    % get the name from dictionary
2  f = evalin('base',slice_name);     % write slice into a temporary variable f
3  f = ...% do some calculatins
4  assignin('base', v, f);            % save result in base workspase
```

Now, the forward- and backprojection operation can be implemented as follows, see algorithm 4.1. The projection data is loaded into the *base* workspace and only the name of the dataset is given to the backprojection function. Inside this function a dictionary for the projection data and the volume slices is created. Then, for each angular view θ, the projection data is accessed from the *base* workspace, processed according to the backprojection algorithm and the result is written back to the *base* workspace using the dictionary for volume slices. It allows for working with twice larger datasets compared

---
Algorithm 4.1: Implementation of backprojection for tomosynthesis using the *base* workspace and a dictionary

---
Input: load $sino(\theta)$, $\theta \in \Theta$ from disc to *base* workspace
Output: Reconstructed volume $slice(z)$, $z \in V$ in *base* workspace

1 **Initialization** : dictionaries for $sino\_name$, $\forall \theta \in \Theta$ and $slice\_name$, $\forall z \in V$;
2 **for** $\theta \in \Theta$ **do**
3   get $sino\_name(\theta)$ from the dictionary;
4   read $sino(\theta)$ from *base* workspace;
5   **for** $z \in Z$ **do**
6           $slice = \mathbf{BP}(sino(\theta))$;
7   get $slice\_name(z)$ from the dictionary;
8   write $slice$ into *base* workspace;

---

to the method with direct passing variables to a function and thus creating copies of those variables.

4.3.3 Memory costs

The size of the projection dataset and the three-dimensionality of the reconstructed tomosynthesis volume require special implementation strategy of any iterative reconstruction algorithm. The implementation has to be adjusted for the optimal usage of available memory and desired speed. There is always a trade-off between the memory costs and the reconstruction time. The memory costs of the SART algorithm will be discussed based on the size of the data acquired using the Siemens Mammomat Inspiration device. The projection dataset contains $N_\theta = 25$ two-dimensional projections. A typical

4.3 Considerations for practical implementation of SART

reconstructed volume contains about $N_{slices} = 40$ reconstructed slices. Both, the projection images and the reconstructed images have size of $N_x \times N_y = 3584 \times 2816$ elements. The memory required to store the tomosynthesis projection dataset is proportional to $N_x \times N_y \times N_\theta$. It requires about 0.9 GB memory if saved as single precision floats. A memory required to store the tomosynthesis reconstructed volume is proportional to $N_x \times N_y \times N_{slices}$. It requires about 1.7 GB memory if saved as single precision floats.

One iteration of the SART algorithm (see algorithm 4.2) requires to keep in memory the measured projection dataset and the reconstructed volume. Moreover, it is required to store the simulated projection dataset (line 8) and the volume with the updating term (line 10), which results in 2.6 GB of additional memory. Pre-calculation of the normalization terms (line 4 and line 5) results in an additional projection dataset for $A_{+,\theta}$ and N_θ volumes of $A_{z,+}$ requires $0.9 + 40 \times 1.7 = 68$ GB, which makes the pre-calculation of $A_{z,+}$ infeasible. If the normalization terms are calculated on-the-fly, they require at least the memory of the reconstructed dataset according to the implementation in algorithm 4.2.

Algorithm 4.2: Straightforward (but infeasible) SART implementation for tomosynthesis

Input: Projection data $sino(\theta)$, $\theta \in \Theta$
Output: Reconstructed volume $slices$, $\forall z \in V$
1 **Initialization** : dictionaries for $sino\_name$, $sino\_upd\_name$ $\forall \theta \in \Theta$;
2 dictionaries for $slice\_name$, upd_term_name $\forall z \in V$;
3 $slices \forall z \in V$;
4 pre-compute $A_{+,\forall \theta} = \mathbf{FP}\left(slice\_ones\left(\forall z\right), \forall \theta\right)$;
5 pre-compute $A_{\forall z,+} = \mathbf{BP}(sino\_ones(\forall \theta), \forall z)$;
6 **for** $\theta \in \Theta$ **do**
7 $\quad$ read $sino\_name$ and $sino\_upd\_name$ for given θ from dictionary;
8 $\quad$ $sino\_upd(\theta) = \mathbf{FP}\left(slice\left(\forall z\right), \theta\right)$;
9 $\quad$ $sino\_upd(\theta) = \left(sino\_upd(\theta) - sino(\theta)\right)/A_{+,\theta}$;
10 $\quad$ $upd\_term(\forall z) = \mathbf{BP}(sino\_upd(\theta), \forall z)$;
11 $\quad$ $slice(\forall z) = slice(\forall z) - (\lambda \cdot upd\_term(\forall z))/A_{\forall z,+}$;

A memory-saving but slower version of SART can be implemented, see (algorithm 4.3). The backprojection operation can be done for each slice location z separately (line 11). After applying the updating term (line 13), the memory is cleared (line 14). After the view θ is processed, the corresponding simulated projection `sino_upd` is also cleared from memory (line 15) because it is not needed anymore.

86   Chapter 4. Iterative image reconstruction

Algorithm 4.3: Memory-saving implementation of SART for tomosynthesis
Input: Projection data $sino(\theta)$, $\theta \in \Theta$
Output: Reconstructed volume $slice(z)$, $z \in V$
1 Initialization : dictionaries for $sino\_name$, $sino\_upd\_name$ $\forall \theta \in \Theta$;
2   dictionaries for $slice\_name$, upd_term_name $\forall z \in V$;
3      $slice$ $\forall z \in V$;
4 for $\theta \in \Theta$ do
5   read $sino\_name$ and $sino\_upd\_name$ for given θ from dictionary;
6      $sino\_upd(\theta) = \mathbf{FP}\left(slice\left(\forall z\right), \theta\right)$;
7      $A_{+,\theta} = \mathbf{FP}\left(slice\_ones\left(\forall z\right), \theta\right)$;
8      $sino\_upd(\theta) = \left(sino\_upd(\theta) - sino(\theta)\right)/A_{+,\theta}$;
9   for $z \in V$ do
10   read upd_term_name and $slice\_name$ for given z from dictionary;
11         $upd\_term(z) = \mathbf{BP}(sino\_upd(\theta), z)$;
12         $A_{z,+} = \mathbf{BP}(sino\_ones(\theta), z)$;
13         $slice(z) = slice(z) - (\lambda \cdot upd\_term(z))/A_{z,+}$;
14   clear $upd\_term(z)$, $A_{z,+}$;
15   clear $sino\_upd(\theta)$, $A_{+,\theta}$;

4.3.4 Computational complexity

The forward- and backprojection steps require the most computational costs in the SART algorithm. Two units for forward- and backprojection computation costs can be defined: a U_{FP} unit and a U_{BP} unit. The U_{FP} unit is a computation cost for a forward projection operation done for one slice and one angular direction. The U_{BP} unit is a computation cost for a backprojection operation done for one slice and one angular direction. Then, one iteration of SART requires the following computational operations. First, the forward projection of the current guess is done, which is equivalent to N_{slices} of U_{FP}. This operation is done N_θ times for each view. Second, the backprojection of an updating term requires the N_{slices} of U_{BP} and this is done N_θ times. This results in total of

$$N_\theta \times N_{slices} \times (U_{FP} + U_{BP}) \qquad (4.27)$$

computational time units for one iteration of SART. The calculation of normalization terms $A_{i,+}$ and $A_{+,j}$ requires additional $N_\theta \times N_{slices} \times (U_{FP} + U_{BP})$ computational time, which doubles the total time for one iteration.

In the distance-driven implementation, used in this work, the forward projection unit U_{FP} is 0.28 sec and the backprojection unit U_{BP} is 0.25 sec. They have been implemented in C++ on Intel(R) Core(TM) i5-2400 CPU @3.10 GHz. In terms of projector costs one iteration of SART is about 530 sec.

4.4 Projection access order for SART

It is known that the order in which the projection data is selected for the iterative scheme has strong influence on the reconstructed image quality and the convergence speed (Guan 1998, Guan 1996, Wu 2008, Kong 2012). In literature, many schemes of presumably optimal access order have been proposed. In this section, several projection access orders found in CT literature will be presented and adapted for the tomosynthesis limited angle case. Additionally, a novel projection access scheme based on total correlation will be presented. All presented schemes will be compared in application to tomosynthesis using a simulation study.

4.4.1 Literature review

It is important to note that in the early papers on various projection acces orders, classical ART is considered, which is a ray-wise algorithm. In this section, simultaneous ART is considered which is view-wise algorithm. Regardless if ART or SART is considered, each discussed approach can be applied *Mutatis mutandis* to either view or rays ordering.

Two simplest methods to choose the projections are the sequential order and the random permutation. It is known that the simple sequential projection access order is not optimal (Mueller 1998a). Another simple choice to generate the projection order is based on a constant angular increment, e.g. 66.0^o or 73.8^o. However the choice of the angular increment is often not motivated and the increment is selected arbitrarily. An example of the motivated choice can be found in the work of T. Kohler. He proposed to use a concept of the *golden ratio* to calculate an optimal angular step (Kohler 2004).

In 1993 G. T. Herman and L. B. Meyer proposed a method based on the prime factors. For PET applications it was superior to the random permutation (Herman 1993). In the literature it is commonly called PND as an abbreviation of the "prime number decomposition". Another permutation scheme based on the prime numbers is the prime number increment (PNI) scheme proposed in 2011 by H. Kong and J. Pan (Kong 2011). In general, this scheme is a scheme with a constant angular step, where the angular increment is selected based on a prime number.

In 1994 H. Q. Guan and R. Gordon proposed a multilevel scheme (MLS) (Guan 1994), which maximizes orthogonality between used projections in each step by minimizing the geometrical correlation between them. Their further studies showed that the

MLS outperforms the sequential order and the random permutation order (Guan 1996, Guan 1998).

Mueller et al. (Mueller 1997) have also recognized that it is important to maximize the angular distance between consecutively used projections. The authors proposed the weighted distance scheme (WDS) to heuristically optimize angular distance between used projections and to take into account the relation to all previously used projections.

To conclude, all existing schemes either use simple methods such as random permutation and fixed angular increment, or minimizes the correlation between used projections based on the angular view information.

4.4.2 Sequential order

In the sequential projection order π_{seq}, all views are accessed sequentially as they have been acquired from the first projection view to the last projection view

$$\pi_{seq}(1) = 1;$$
$$\pi_{seq}(i) = \pi_{seq}(i-1) + 1, \ i = 2 : N_\gamma; \quad (4.28)$$
$$\pi_{seq} = [1, 2, 3, ..., N_\gamma].$$

If a limited number of projections $N_{proj} = 25$ is measured, the sequential projection access order can be easily created. Due to non-symmetry of the acquired angular data with respect to the 360°, the starting angle might also be an important parameter.

The sequential access is known to introduce artifacts to the reconstructed images (Mueller 1998a). Almost no new information is introduced to the reconstructed image by the next used projection. A large geometrical correlation between the neighbor views slows the recovery of high frequency and, therefore, slows down the convergence rate. As it was concluded by Guan et al. (Guan 1998), algebraic reconstruction with sequential projection access order cannot reconstruct objects uniformly and symmetrically.

4.4.3 Random permutation

In the random permutation order π_{rnd} each next view is selected arbitrarily, independent of the total number of views. In MATLAB® the random projection access order can be produced e.g. using the randperm build-in function. M.C.A. van Dijke concluded in 1992 in his PhD thesis (van Dijke 1992) that the random permutation is the best method among all methods which he tried. However, methods which are more controllable and predictable are preferred (Mueller 1998a).

4.4 Projection access order for SART

In equation 4.29 an example of the random permutation order for $N_{proj} = 25$ is shown.

$$N_{proj} = 25;$$
$$\pi_{rnd} = [19, 9, 21, 22, 3, 17, 20, 25, 15, 4, 1, 16, ... \quad (4.29)$$
$$8, 12, 24, 18, 14, 13, 6, 2, 23, 10, 5, 11, 7].$$

4.4.4 Golden ratio

The projection access order based on the golden ratio π_{golden} (Kohler 2004) uses a constant angular increment. The choice of the increment is this scheme is inspired by the way many plants position their new leafs, a so-called *golden ratio*

$$g = \frac{\sqrt{5} - 1}{2} \approx 0.618. \quad (4.30)$$

A new leaf is placed at an angle such that there is minimum overlap to prevent unwanted shadow. The golden angular increment is $360°g$ which is approximately $222.5°$. T. Kohler considered a parallel beam geometry, therefore his golden angle is $180°g$ which is equal $111.24°$. According to the results presented in his paper, this method performs better than the random scheme and better or equal the prime number decomposition. In addition, it is easier to implement than PND or WDS.

Extending this concept, a golden angle for the limited angle tomosynthesis geometry can be found. If the total angular range is $50°$, then the golden angle increment is $50°g$ which is equal to $30.9°$. In equation 4.31 an example of the golden ratio sequence for $N_{proj} = 25$ and the golden angular increment $\Delta\theta_{golden} = 30.9°$ is shown. If by applying this angular increment the resulting value of angle is outside the valid region, the value is wrapped around by using the modulo operation. Calculations are done based on the angular information taken from the DICOM header of the real data acquired using the Siemens Mammomat Inspiration scanner.

$$N_{proj} = 25; \quad \Delta\theta_{golden} = 30.9°$$
$$\pi_{golden} = [17, 7, 23, 13, 3, 19, 9, 1, 15, 5, 22, 12, ... \quad (4.31)$$
$$2, 18, 8, 24, 14, 4, 20, 10, 16, 6, 25, 21, 11].$$

4.4.5 Prime numbers decomposition (PND)

The permutation access order based on the prime numbers decomposition was proposed by G. T. Herman and L. B. Meyer in 1993 (Herman 1993). In the following explanation of the method, the notation from the original paper with some modifications is used.

Modifications are needed because the notations of the original paper might lead to confusion. The symbol p and symbol u are used in their paper for different entities throughout the text of the original paper.

Let K be the number of measured views. The prime number permutation $\tau(k) = k'$ returns a new ordered sequence of projections k'. Let $\{p_i\}_{i=1}^{U}$ be an ascendingly sorted set of prime factors of K with the total number U

$$\begin{aligned} K &= p_1 \times p_2 \times ... \times p_U; \\ p_1 &\leq p_2 \leq ... \leq p_U. \end{aligned} \quad (4.32)$$

Let us consider a set T of U-dimensional non-negative vectors t, which i-th component t_i is non-negative and less than the corresponding prime factor p_i

$$\begin{aligned} t &= \{t_1, t_2, ..., t_U\} \in Z^* = \{0\} \cup Z^+ \\ 0 &\leq t_i \leq p_i. \end{aligned} \quad (4.33)$$

Let us define a mapping

$$\begin{aligned} \tau: \ &[0, P) \to T \\ &k \mapsto t, \end{aligned} \quad (4.34)$$

which maps an integer k into a vector t from T. The mapping τ is defined recursively. The initialization for $k = 0$ is $\tau(0) = t^{(0)} = [0, 0, ...0]$. Then for each $k > 0$ an integer s is assigned. The integer s is the smallest integer such that $t_s^{(k-1)} \neq p_s - 1$, where p_s is the s-th prime factor of K. The i-th component of the vector $t_i^{(k)}$ is

$$\tau(k): t_i^{(k)} = \begin{cases} 0, & \text{if } 1 \leq i < s; \\ t_i^{(k-1)} + 1, & \text{if } i = s; \\ t_i^{(k-1)}, & \text{if } s < i \leq U, \end{cases} \quad (4.35)$$

where i is from 1 to U.

Lets define another mapping

$$\begin{aligned} \upsilon: \ &T \to [0, P) \\ &t \mapsto k', \end{aligned} \quad (4.36)$$

4.4 Projection access order for SART

which maps a vector t from T onto an integer k'.

$$\begin{aligned}v(t) =\ & (p_U \times p_{U-1} \times ... \times p_2 \times t_1) \\ & +(p_U \times p_{U-1} \times ... \times p_3 \times t_2) \\ & +... \\ & +(p_U \times t_{U-1}) + t_U;\end{aligned} \qquad (4.37)$$

Then, the permutation scheme $\pi(k)$ is defined as $v(t^{(k)}) = v(\tau(k))$, see algorithm 4.4.

Algorithm 4.4: Prime numbers decomposition permutation
Input: number of measured views K
Output: new permutation order k'
// $\tau(k)$ and $v(t)$ are calculated based on the prime factors of K ;
1 find $p_1...p_U$ prime factors of K;
2 $k = 0$; $t^{(0)} = [0, 0, ...0]$; // an initialization
3 **for** $k = 1 : K - 1$ **do**
4   find a smalles integer s;
5      $t^{(k)} = \tau(k)$; // assign a vector t_k, see equation 4.35
6      $k' = \pi(k) = v(t^{(k)}) = v(\tau(k))$; // find a new k' based on the vector t_k, see equation 4.37

In the case of tomosynthesis geometry the number of projections is $N_p = 25$, its prime factors are $p_1 = p_2 = 2$, $U = 2$. The set of vectors is $T = \{t_1, t_2\}$; with $t_1 : 0 \le t_1 \le 4$; $t_2 : 0 \le t_2 \le 4$. Then, equation 4.35 can be rewritten as

$$\begin{aligned} t_1^{(k)} &= \begin{cases} t_1^{(k-1)} + 1, & \text{if } s = 1 \\ 0, & \text{if } s = 2 \end{cases} \\ t_2^{(k)} &= \begin{cases} t_2^{(k-1)} =, & \text{if } s = 1 \\ t_2^{(k-1)} + 1, & \text{if } s = 2 \end{cases} \end{aligned} \qquad (4.38)$$

Additionally, the equation 4.37 is simplified to $v(t) = p_2 \times t_1 + t_2$. The details of producing the permutation for $N_p = 25$ can be found in Table 4.1.

Table 4.1: Prime numbers decomposition permutation access for $N_p = 25$.
Prime factors are $p_1 = p_2 = 5$; $U = 2$; $v(t) = p_2 \times t_1 + t_2$, see equation 4.37.

k	s	$t = \tau(k, s)$	$\pi(k)_{PND} = v(t)$
0	-	(0, 0)	0
1	1	(1, 0)	5
2	1	(2, 0)	10
3	1	(3, 0)	15
4	1	(4, 0)	20
5	2	(0, 1)	1
6	1	(1, 1)	6
7	1	(2, 1)	11
8	1	(3, 1)	16
9	1	(4, 1)	21
10	2	(0, 2)	2
11	1	(1, 2)	7
12	1	(2, 2)	12
13	1	(3, 2)	17
14	1	(4, 2)	22
15	2	(0, 3)	3
16	1	(1, 3)	8
17	1	(2, 3)	13
18	1	(3, 3)	18
19	1	(4, 3)	23
20	2	(0, 4)	4
21	1	(1, 4)	9
22	1	(2, 4)	14
23	1	(3, 4)	17
24	1	(4, 4)	24

4.4 Projection access order for SART

4.4.6 Prime number increment (PNI)

Another algorithm which involves prime numbers was proposed by Kong (Kong 2011). It works as the opposite of the PND algorithm. A prime number P is used, which is *not* a prime factor of the number of measured views K

$$\begin{aligned}\pi(0) &= 0; \\ \pi(k) &= (\pi(k-1) + P) \mod (K), \ 1 \leq k \leq K-1.\end{aligned} \quad (4.39)$$

The method is simple, however no advice how to choose an optimal P for given K is given in the publication. In the equation 4.40 an example of a PNI sequence for $K = 25$ and $P = 7$ is shown.

$$\begin{aligned}K &= 25; \ P = 7 \\ \pi_{PNI} &= [0, 7, 14, 21, 3, 10, 17, 24, 6, 13, ... \\ &\quad 20, 2, 9, 16, 23, 5, 12, 19, 1, 8, 15, 22, 4, 11, 18]\end{aligned} \quad (4.40)$$

4.4.7 Multilevel scheme (MLS)

According to H. Q. Guan and R. Gordon (Guan 1994), a high geometrical correlations between used projections makes the algebraic reconstruction slow in convergence. The projections, which are $90°$ apart, have the minimum correlation and introduce the most independent information into reconstruction. Based on the geometrical correlations of projections the MLS scheme was developed. The MLS orders the projections to keep the geometrical correlation to the last used projection and to the other already accessed projections minimal.

The MLS organizes all projections in certain levels, denoted by l. It starts with view number 0 ($0°$) and view $N_{proj}/2$ ($90°$) as they have the maximum orthogonality. The second level includes projections $N_{proj}/4$ ($45°$) and $3N_{proj}/4$ ($0°$). Starting from the second level, the *index* is generated based on the previously used indexes at all previous layers by adding a $factor$ $N_{proj}/2^l$, see Table 4.2. The mapping from the *index* to the projection view is done by $\pi_{MLS} = index \times factor(l)$. Apparently, the number of used projections in each level is equal to the $2^{(l-1)}$ or the total number of the used in the previous levels projection.

For example, the third level includes four projections because two projections were used in the first level and two projections were used in the second level. The *index* sequence of the third level can be calculated as

$$\begin{aligned}\pi_{MLS}(l=3) : \ &[0 \times \tfrac{N_{proj}}{2}, \ 1 \times \tfrac{N_{proj}}{2}, \ 1 \times \tfrac{N_{proj}}{4}, \ 3 \times \tfrac{N_{proj}}{4}] + \tfrac{N_{proj}}{2^3} = \\ &1 \times \tfrac{N_{proj}}{8}, \ 5 \times \tfrac{N_{proj}}{8}, \ 3 \times \tfrac{N_{proj}}{8}, \ 7 \times \tfrac{N_{proj}}{8}.\end{aligned} \quad (4.41)$$

Table 4.2: Access order for MLS.

$level\ l$	length $2^{(l-1)}$	$factor$	$index$
1	2	$\times N_{proj}/2$	$[0, 1]$
2	2	$\times N_{proj}/4$	$[1, 3]$
3	4	$\times N_{proj}/8$	$[1, 5, 3, 7]$
4	8	$\times N_{proj}/16$	$[1, 9, 5, 13, 3, 11, 7, 15]$
5	16	$\times N_{proj}/32$	$[1, 17, 9, 25,...]$

The MLS requires the number of projections N_{proj} be a power of two. If N_{proj} is not a power of two, additional modifications including rounding the values to integers and rejection of already used values, since the same numbers may appear more than once before all projections are processed.

In equation 4.42 an example of an MLS sequence for $N_{proj} = 25$ is shown. The repeating numbers have been ignored when creating this sequence.

$$N_{proj} = 25;$$
$$\pi_{MLS} = \begin{array}{l}[1,\ 13,\ 7,\ 19,\ 4,\ 16,\ 10,\ 22,\ 2,\ 15,\ 8,\ 21,... \\ 5,\ 18,\ 11,\ 24,\ 14,\ 20,\ 17,\ 23,\ 3,\ 9,\ 6,\ 12,\ 25].\end{array} \quad (4.42)$$

4.4.8 Weighted distance scheme (WDS)

The WDS scheme proposed by Mueller et.al. (Mueller 1997) also minimizes the correlation between used projections. This scheme consists of two phases, the initial filling and the update phase. In the initial phase all projections are ordered as a circular queue Θ. In the update phase old values in Θ are overwritten by new values. New values are chosen based on the minimization of the weighted mean of *repulsive forces* between projections and on the minimization of the weighted standard deviation of the distances between projections. The WDS is not included into the current study.

4.4.9 Data-based minimum total correlation order

All projection access orders share a similar concept that the next applied projection should bring as much new information as possible, i.e. to have small correlation with the previously used projection(s). All schemes discussed above minimize the correlation between selected projections by using only the information at which angle the projection was measured. In this subsection, the minimum-correlation approach for construction

4.4 Projection access order for SART

the projection access order will be extended by using the object-related information instead of only the angular information.

The main idea is to calculate explicitly the correlation between each projection images and to find such projection order which results in the total minimal correction. Correlation between each projection image can be represented as a matrix of $N_{proj} \times N_{proj}$ size.

A correlation coefficient $c_{j_1 j_2}$ between two projection images p_{j_1} and p_{j_2} is given by

$$c_{j_1 j_2} = \frac{\sum\limits_{i}^{M} \left(p_{j_1} - \bar{p}_{j_1}\right)\left(p_{j_2} - \bar{p}_{j_2}\right)}{\sqrt{\sum\limits_{i}^{M} \left(p_{j_1} - \bar{p}_{j_1}\right)^2 \sum\limits_{i}^{M} \left(p_{j_2} - \bar{p}_{j_2}\right)^2}}, \qquad (4.43)$$

where $\bar{p}_{j_1}$ and $\bar{p}_{j_2}$ are the mean values. Then, a path in the matrix c is searched, which leads to the minimum correlation. A simple nearest-neighbor approach can be used as well as more sophisticated minimum path finding algorithms (Kiencke 2013a).

4.4.10 Simulation results

A comparison of several projection access orders for SART in application to tomosynthesis geometry has been done. The sequential, random, PND, MLS, golden ratio and minimum correlation approaches have been considered. The noiseless projection data have been simulated using the geometry of the Siemens Mammomat Inspiration device. A phantom contains the concentric small sphere of 5 mm radius and a cylinder with 50 mm radius.

To construct the minimum correlation order, first, the correlation matrix has been calculated, see Fig. 4.1a. Then, the path has been found using the nearest neighbor approach, also known as a greedy approach (Kiencke 2013b). The obtained order is visually shown in Fig. 4.1b.

A plot showing the costs to move from the current projection to the next projection is shown in Fig. 4.2. It can be seen, that the sequential order has the highest costs for each step. Some pattern can be recognized in the curve for PND. The costs for random, MLS and golden orders are scattered. The costs for the minimum correlation order increase with each step, starting from small values and ending up with the costs similar to the sequential approach. This behavior can be explained by the used path search algorithm. The nearest neighbor method finds some paths, but it is not guaranteed that it is the absolute minimum path. The comparison of the total path length is shown in Table 4.3, first row. Among all, the minimum correlation order has the shortest path length.

Chapter 4. Iterative image reconstruction

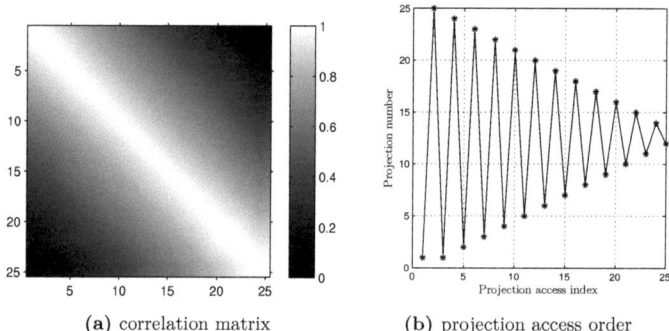

(a) correlation matrix   (b) projection access order

Figure 4.1: The minimum correlation access order. (a) correlation matrix and (b) visualization of the projection access order.

Table 4.3: Total length of the path in the correlation matrix and the total lengths including the history of the previously used projections (one element).

order	seq	rnd	pnd	mls	golden	mincorr
length	23.57	16.37	16.19	13.49	12.16	**10.52**
length with history	44.36	29.93	28.00	30.47	**26.01**	33.13

Six volumes have been reconstructed using SART with zero-valued initial guess, 20 iterations and the abovementioned projection access orders. Reconstructed volumes have been quantitatively compared with the original phantom using the normalized root mean squared error (NRMSE). The NRMSE versus the iteration number is shown in Fig. 4.3. Although, the minimum correlation order shows promising behavior in the first iterations, the reconstruction process seems to stuck at a local minimum. It can be explained by the fact, that the nearest-neighbor algorithm does not take into account any history, which, in turn, might be important. If the currently used projection p_j has a small correlation to the previously used projection p_{j-1}, but has a large correlation to the projection p_{j-2}, it will slow the convergence rate. The total path with taking into account the history (one element, see Table 4.3, second row) shows that the minimum correlation order, indeed, does not produce the smallest correlation. Therefore, better algorithms for path searching, which include the history of the previously used projections, are required (Kiencke 2013a).

4.4 Projection access order for SART

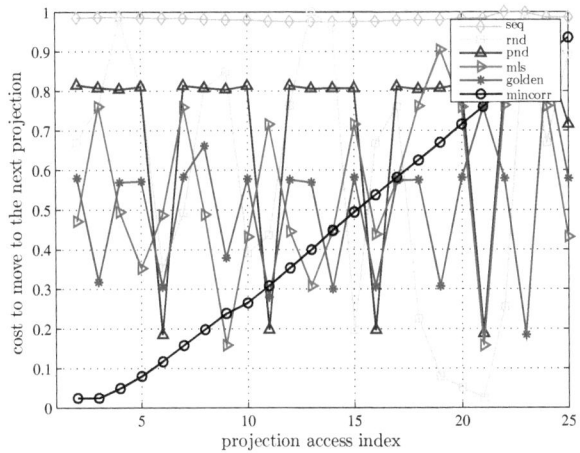

Figure 4.2: Costs to move to next projection for different projection access orders.

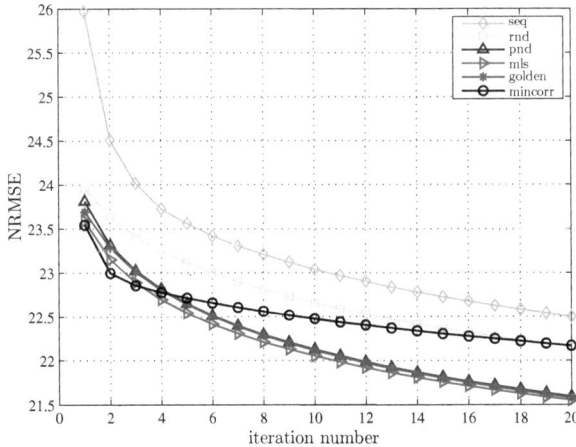

Figure 4.3: The NRMSE between the reconstructed volume and the reference volume for different projection access orders.

Chapter 5

Backprojected space in image reconstruction

Contents

5.1	Theory of backprojected space	100
5.2	A weighting scheme based on dissimilarity	106
5.3	Non-linear backprojection ωBP for tomosynthesis	111
5.4	Weighted ωSART for tomosynthesis	120
5.5	Weighted ωSART for metal artifact reduction in CT	125
5.6	Interpolation in BP-space for metal artifact reduction in CT	133

Backprojected space (BP-space) is a sinusoid-like decomposition of the projection data. The BP-space is a four-dimensional space which connects the projection space and the image space. It gives a new possibility to represent the backprojection operator and to control it with more flexibility. It allows for introducing non-linear weighting coefficients in the backprojection operator to control the contribution of each projection value into each voxel individually. It can be used when the projection data are incomplete or inconsistent to reduce the influence of incompleteness/inconsistency of the data. Additionally, the BP-space offers a new methodology to follow the sinogram flow without any segmentation or registration. This can be used for data interpolation.

This chapter includes a literature review of the *stackgram* representation which is the BP-space in the two-dimensional case. The BP-space will be introduced as generalization of the stackgram approach for the third dimension and the properties of the BP-space will be discussed. Then, a motivation for a non-linear weighted backprojection for tomosynthesis will be given (Levakhina 2012a). The weighting scheme will be discussed and reconstruction results will be shown. Afterwards, the weighting scheme will be extended for the usage in the SART algorithm with some results (Levakhina 2012c, Levakhina 2013b). As the last part of this chapter, unpublished work regarding metal artifact reduction in CT using ωSART and interpolation in BP-space will be presented.

5.1 Theory of backprojected space

In this section, the literature review and the properties of the stackgram representation and the BP-space will be presented. To the best of our knowledge, the detailed properties of the planes of different orientation in the stackgram and BP-space have never been discussed in literature before.

5.1.1 Stackgram representation in literature

A concept of the *stackgram* representation has been introduced by A. Happonen in 2002 and summarized in 2005 in his PhD dissertation (Happonen 2005a). The stackgram is an intermediate three-dimensional domain between the two-dimensional image domain and the two-dimensional sinogram domain. First, it has been proposed for sinogram denoising of PET data. Data processing along sinusoidal curves has a potential (Happonen 2002, Krestyannikov 2004a) and a superiority (Happonen 2005b) to other sinogram filtering methods for this imaging modality. An application of stackgram denoising to the attenuation-corrected PET data can be found in (Krestyannikov 2004b). A study on exact formulation of filters can be found in (Peltonen 2010). A similar technique has also been proposed for noise reduction of low-dose CT data (Happonen 2007b). The stackgram filtering was also used for denoising of SPECT data (Happonen 2007a). A method based on a similarity comparison within the neighborhood of locus-signals in stackgram has been proposed for alignment of tomographic data in dynamic PET (Happonen 2003, Happonen 2004, Kostopoulos 2006). Another application of the stackgram, which can be found in literature, is an extrapolation of limited-angle data (Happonen 2005c). A similar sinogram decomposition approach has been proposed by A. Zamyatin for restoration of the truncated data (Zamyatin 2007). J. Caramelo (Caramelo 2005) proposed a reconstruction method based on the sinogram decomposition into single sinogram curves, which has similarities to the stackgram approach. Another paper was published in 2012 which proposes an inpainting based on sinusiod-like curve decomposition and uses an eigenvector-guided interpolation (Li 2012).

5.1 Theory of backprojected space

In the next subsection the stackgram representation will be extended to the fully three-dimensional case and its properties will be discussed in details.

5.1.2 Properties of BP-space

The BP-space is a generalization of the stackgram representation for dimensions higher than two. Here, it will be addressed for the three-dimensional geometry. The BP-space is constructed using a modified backprojection operator such that the summation (integration in continuous case) is replaced by a *stack* operator.

In case of two-dimensional parallel-beam geometry, the stack operator **S** is defined by equation 5.1. Here, D denotes the object support.

$$h(x, y, \theta) = \mathbf{S}p(\xi, \theta) = p(x\cos\theta + y\sin\theta, \theta), \\ (x, y) \in D \subset R^2, \ \theta \in [0, 2\pi) \tag{5.1}$$

It maps each i-th sinogram row $p(\xi, \theta = \theta_i)$ into the i-th plane in the BP-space for all view directions, i.e. $\forall i = 1 : N_\theta$. The resulting two-dimensional images are stacked along the θ direction for each angular view. This results in a three-dimensional matrix $h(x, y, \theta)$. To go back from the BP-space to the sinogram, a modified FP-operator (equation 5.2) can be used, such that each plane of the BP-volume $h(x, y, \theta)$ is forward projected according the corresponding angular view θ and results in a sinogram row. Rows are combined together into a sinogram.

$$p(\xi)_\theta = FP_\theta [h(x, y, \theta)] = \iint_D h_\theta(x, y) \delta(x\cos\theta + y\sin\theta - \xi) \, dxdy, \ \theta \in [0, 2\pi) \tag{5.2}$$

In practical applications only coordinates within the object support D are considered and the result is referred to as a *BP-volume*. An example of the BP-volume representation of an object consisting of a point-like feature is presented in Fig. 5.1a. It looks like a stack of spokes with a single spoke lying in each theta-plane. Each spoke is rotated by θ in xy-plane. All of them form a spiral pattern in the volume. In Fig. 5.1b an example of the BP-volume of a lung slice from the XCAT phantom (Segars 010) is shown. Each point in the image produces spiral spokes. Their superposition results in the complex-looking structures.

An xy-plane of the BP-volume contains a set of lines, also known as ridge functions (Logan 1975, Candes 1999). The lines in each plane have the corresponding θ orientation. If no post-processing is applied, the information is redundant along each line. Examples of the xy-plane of a PB-volume of a point-like object and a PB-volume of the lung slice from the XCAT phantom are shown in Fig. 5.2.

Chapter 5. Backprojected space

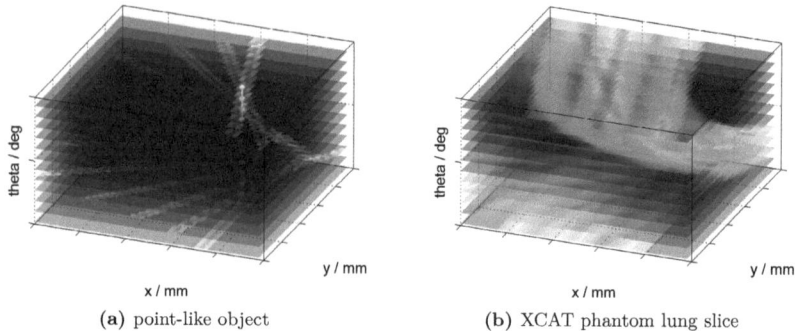

(a) point-like object   (b) XCAT phantom lung slice

Figure 5.1: Backprojected volume of a point-like object (a) and a lung slice from XCAT phantom (b). A Matlab *jet* colormap is used for better visualization, blue color corresponds to zero gray value.

A vector $h_{x_0 y_0}(\theta)$ contains all backprojected values, which contribute from all angular views to the selected pixel (x_0, y_0). In the case when the image contains only one point-like feature located in the pixel (x_0, y_0), the corresponding projection data $p(\xi, \theta)$ will contain exactly one sinusoid-like curve. Lets consider an object with a set of point-like features distributed along the line $x = x_0$ and focus on a feature located at (x_0, y_0), see Fig. 5.3a.

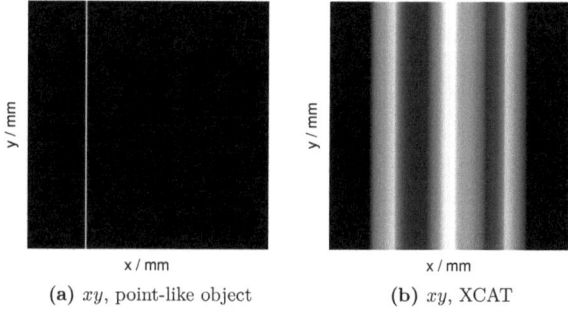

(a) xy, point-like object   (b) xy, XCAT

Figure 5.2: An example of xy-planes with orientation angle $\theta = 0^o$. The information is redundant along each line (ridge functions).

5.1 Theory of backprojected space 103

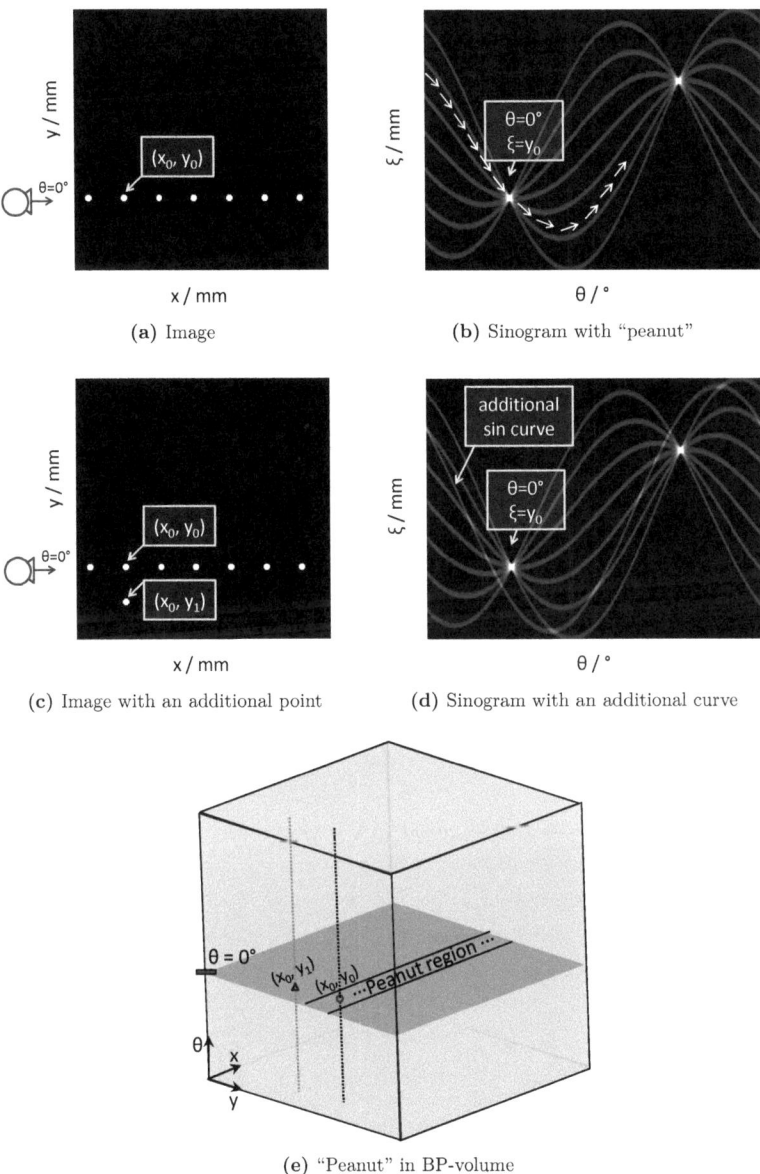

(a) Image

(b) Sinogram with "peanut"

(c) Image with an additional point

(d) Sinogram with an additional curve

(e) "Peanut" in BP-volume

Figure 5.3: Connection between the line in the image domain, the "peanut"-shape in the sinogram domain and the corresponding coordinates in the BP-volume.

The corresponding sinogram contains a peanut-like bunch of sinusoidal curves. The curve produced by the feature (x_0, y_0) is marked by arrows. All sinusoidal curves cross in one point, where the view direction θ_0 is parallel to the x-axis, and the detector coordinate $\xi = y_0$, see Fig. 5.3b. Describing the situation vice-versa: the points, which contribute to the sinogram value $(\xi = y_0, \theta = 0)$ lie within a peanut-shape in the sinogram. Using the concept of the BP-volume, we can easily move along each of the sinusoids within the peanut shape region. If we move in the BP-volume along a θ-vector located at (x_0, y_0), we are moving along the sinusoidal curve produced by the corresponding pixel, see Fig. 5.3e.

If we select $\theta = 0$, fix the coordinate x_0, move along ridge line, select an arbitrary value y and then move along the θ-vector, it is equivalent to the moving along a curve in the peanut bunch. At the same time, an additional point (x_0, y_1) (see Fig. 5.3c) will produce a curve, which does not belong to the abovementioned peanut-bunch (Fig. 5.3d) and can be accessed via a θ-vector located at (x_0, y_1), see triangle marker in Fig. 5.3e.

The $x\theta$- and $y\theta$-planes represent a set of θ-vectors for the selected x or y coordinate, respectively. Those planes contain information within peanut-like regions in the sinogram. Consider a BP-volume for an image, which contains one point-like object. Then all planes drawn through an arbitrary coordinate will contain some parts of the sinusoidal trace, see Fig. 5.4a and Fig. 5.4c. If a plane passes through the coordinate of the feature, it will contain a "singularity", i.e. there will be a θ-vector present, where all entries contain the values of the sinusoid curve associated with this feature, see Fig. 5.4b and Fig. 5.4d. The same behavior is observed when a realistic image is considered, e.g. a lung slice from the XCAT phantom. Hoverer, if no high-absorption features are present, "singularities" are not obviously visible, see Fig. 5.4e - Fig. 5.4h. Both, $x\theta$- and $y\theta$-planes can be seen as a decomposition of a profile of the point-spread function (PSF), or more precisely the contribution into PSF before the summation operation is done.

In case of the three-dimensional imaging geometry, a similar operator $\mathbf{S^{3D}}$ can be defined, resulting in the four-dimensional backprojected space $h^{3D}(x, y, z, \theta)$

$$h^{3D}(x, y, z, \theta) = \mathbf{S^{3D}} p(u, v, \theta). \qquad (5.3)$$

The angle θ describes the X-ray tube position and (u, v) are coordinates of a point on the detector. The parameters (u, v) and θ are defined by the acquisition geometry. In contrast to the two-dimensional parallel beam geometry where the $\mathbf{S}$ can easily be defined (equation 5.1), in the three-dimensional case $\mathbf{S^{3D}}$ cannot be described as a general expression because it depends on the specific geometry of the scanner.

5.1 Theory of backprojected space

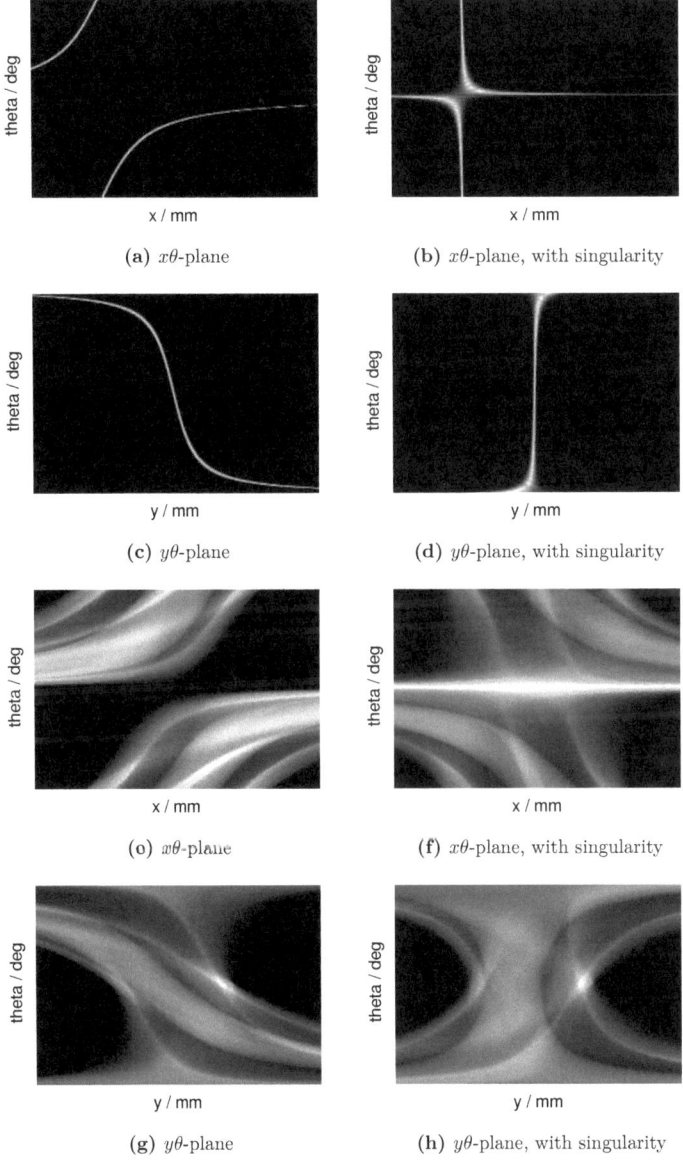

(a) $x\theta$-plane

(b) $x\theta$-plane, with singularity

(c) $y\theta$-plane

(d) $y\theta$-plane, with singularity

(e) $x\theta$-plane

(f) $x\theta$-plane, with singularity

(g) $y\theta$-plane

(h) $y\theta$-plane, with singularity

Figure 5.4: A visualization of the θx- and θy-planes of a BP-volume. (a)-(d) correspond to a point-like object and (e)-(h) correspond to a slice from the XCAT phantom.

5.2 A weighting scheme based on dissimilarity

In this section a weighting scheme for tomosynthesis is presented. The weighting is designed to reduce the out-of-focus tomosynthesis artifacts produced by the high-attenuation features. The section is based on (Levakhina 2012a) and (Levakhina 2012c).

5.2.1 Motivation for weighting: tomosynthesis blur formation

The motivation for the proposed weighting scheme arises directly from the blur formation principle in conventional tomosynthesis (conventional tomography), see e.g. (Ziedses des Plantes 1932). In the conventional tomography one complete movement of an X-ray sensitive film and an X-ray source resulted in one slice through the body. During an acquisition the film and the tube were moved continuously in opposite directions across the patient. This brings into focus only one one plane, which contains the iso-center of motion, The structures located in this plane appeared sharp. At the same time, all other structures located above and below this plane appeared blurred due to the inherent averaging process. In order to visualize another plane, the iso-center was moved to the new position and a new acquisition was performed. The integration process was done directly on the sensitive film. As it has been mentioned in the Chapter 2, a great step forward was an idea to acquire the data as a sequence of analogue projection images, rather than as one single analogue image. An infinite number of slices through the object can be produced if the sequence of projection images is measured, see e.g. (Miller 1971).

A modern DT acquisition results in a sequence of digital images. The simple backprojection operation includes the backprojection of each digital image and accumulating the result as a volume. The BP-space representation offers a possibility to process the contribution from each projection separately. In other words, the simple backprojection can be subdivided into two steps: the backprojecting according to the geometry and the summation of the result in each voxel. Consider a two-dimensional tomosynthesis example with three measurement view directions. The measurement process is schematically shown in Fig. 5.5a and the simple backprojection is shown in Fig. 5.5b. The selected plane of interest with a triangle feature is marked by a dashed line. A circle feature, which is located below the plane of interest, will contribute to the blur. A detailed look into the backprojecting step is shown in Fig. 5.5c. It can be seen, that after the backprojection the triangle feature appears sharp in the selected plane. At the same time, the circle feature appears as multiple ghosting copies, i.e. it produces blur in this plane. This happens because the triangle belongs to the selected plane and the corresponding projection values are always backprojected onto the correct location (see Fig. 5.5c, voxel x_1). The circle does not belong to the selected plane and, therefore,

5.2 A weighting scheme based on dissimilarity

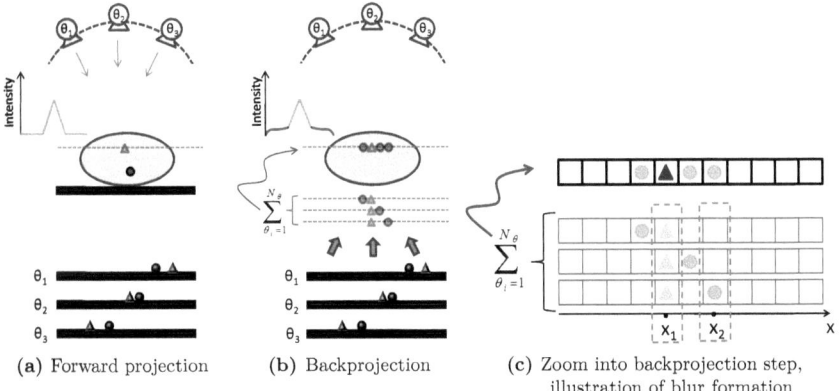

(a) Forward projection   (b) Backprojection   (c) Zoom into backprojection step, illustration of blur formation

Figure 5.5: The principle of blur formation in tomosynthesis and a motivation for the weighted backprojection.

it is always backprojected onto the wrong location (see Fig. 5.5c, e.g. voxel x_2). Thus, the blurring principle can be formulated as follows: because each measured projection contains overlapping details of the object, structures lying in the plane of interest will coincide in each backprojection after the backprojection step (Fig. 5.5c, voxel x_1). At the same time, structures located outside the plane will not coincide and thus will contribute to an undesired blur (Fig. 5.5c, voxel x_2).

Taking into account the information if a particular feature belongs to the plane of interest or not, it is possible to introduce the weighting coefficients into backprojection. One has to identify rays which contain contributions from out-of-focus structures and assign a small weight for the selected plane. This can be done by a comparison of all values, which contributes to the selected voxel. In Fig. 5.5c, all contributing values to the voxel x_1 are identical. Therefore, no weighting is needed. At the same time, in the voxel x_2 the contribution from the circle feature is noticeably larger than other values. Therefore, a small weight has to be assigned to this contribution. This way, it is possible to suppress contributions to artifacts from high-attenuation structures when they are backprojected onto wrong geometrical locations and to preserve the in-focus structures. As a result, the contributions to artifacts are suppressed and the in-focus structures are preserved.

5.2.2 Dissimilarity degree

In case when a volume contains only one point-like feature located in the voxel (x', y', z'), the corresponding projection data $p(u, v, \theta)$ will contain exactly one three-dimensional sinusoid-like curve. All entries of the corresponding θ-vector $h_{x'y'}(\theta)$ in the backprojected space will have the same value. In the case when a volume contains two point-like features (Fig. 5.6a), the corresponding projection data will contain two crossing sinusoid-like curves (Fig. 5.6b) and the entries of the selected θ-vector in BP-space will contain not only one constant value (star-shaped markers), but also a few outliers, which correspond to those angular views, when two sinusoids are crossed (Fig. 5.6c).

In the case of medically relevant objects, the projection data contain a large number of overlapping sinusoid-like curves and each entry in the θ-vector might have a different value. When an object contains a high-absorption feature in the voxel (x', y', z'), the sinusoidal-like curves produced by this feature will cross the sinusoidal-like curve produced by the (x, y, z) point. The values on the crossing location will be relatively large compared to the rest of the values. Here, one can assign a *dissimilarity degree* to each value in the θ-vector and identify outliers. The dissimilarity degree is a positive value less or equal than one, describing how far away the current value is from the whole ensemble of values. The outliers come from the high-absorption features and potentially produce artifacts. At the same time, the sinusoid-like curve produced by the voxel (x', y', z') will contain no outliers because all values will be similar with the same relative large value. This allows for introducing the spatially-depended adaptive weighting coefficients to suppress non-similar values based on their dissimilarity degree.

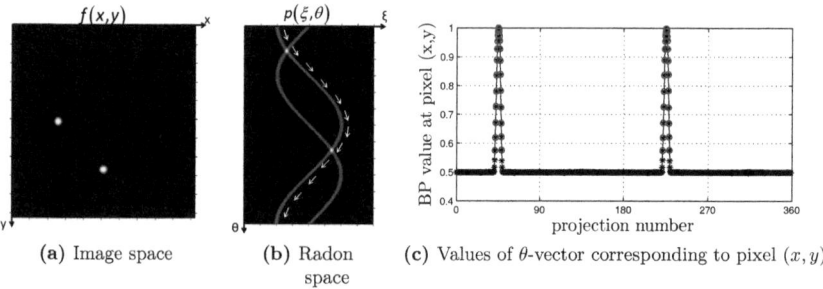

(a) Image space   (b) Radon space   (c) Values of θ-vector corresponding to pixel (x, y)

Figure 5.6: (a) image with two point features; (b) sinusoid curve corresponding to pixel (x, y) is marked with arrows; (c) values along the sinusoidal curve or θ-vector in the BP-space (star-markers), outliers are marked by circle-shaped markers.

5.2.3 Weighting scheme

The idea of the weighting scheme has been inspired by the following two algorithms. The first algorithm is the extreme-value decoding, where only projection data with minimum or maximum value are used for the reconstruction (Haaker 1985a, Haaker 1985b, Stiel 1993). The second algorithm is the voting strategy for statistical reconstruction in breast tomosynthesis when projections with too high value are detected, segmented and completely rejected (Wu 2006).

The dissimilarity degree can be used to correct for a too large contribution in the simple backprojection when a backprojected contribution is calculated using a large sinogram value and is back-distributed onto the wrong geometrical location. The proposed algorithm does not use any segmentation or tissue-classification steps. The detection and correction of values is done automatically.

First, for each voxel (x, y, z) within the set of volume points V the BP-volume $h(x, y, z, \theta)$ is constructed using a stack operator $\mathbf{S^{3D}}$

$$\forall\ (x, y, z) \in V,\ \forall\ \theta \in \Gamma : h^{3D}(x, y, z, \theta) = \mathbf{S}p(u, v, \theta). \quad (5.4)$$

The stack operator is defined by the geometry and imaging parameters of the device. A schematic representation of θ-vector is shown in Fig. 5.7a. Then, for each element of each θ-vector a dissimilarity degree $d_{xyz}(\theta)$ is assigned

$$d_{xyz}(\theta) = \frac{|h_{xyz}(\theta) - M_{xyz}|}{range},\ d \in (0, 1). \quad (5.5)$$

The dissimilarity is defined as the absolute difference to a reference value M_{xyz}, normalized to the $range$ of the projection data, excluding the high-attenuation features. The $range$ is defined as the difference between the largest and the smallest possible value. The reference value can be chosen as the mean, minimum, median, pairwise distinction or other values based on the statistics of the θ-vector. The largest dissimilarity coefficient value is defined to be equal one. If some dissimilarity coefficient are larger than one, they should be limited to one. It is important to note that the reference value M is not considered as the expected solution and the algorithm does not converge to this value. A schematic representation of the dissimilarity is shown in Fig. 5.7b. The reference value M is shown as a dashed line and potential outliers with large distance to the reference value are marked by triangle markers.

The weighting coefficients $\omega_{xyz}(\theta)$ are calculated as a function of dissimilarity. It is reasonable to assume the relation between the dissimilarity and the weighting coefficients

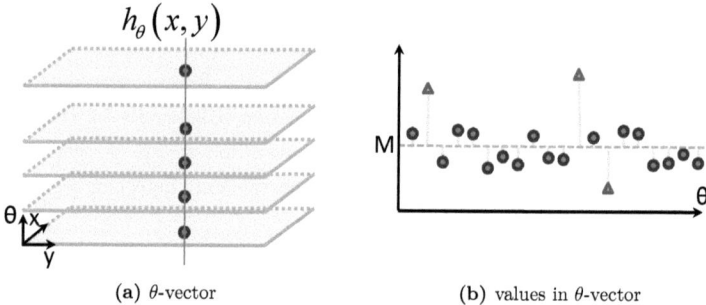

(a) θ-vector (b) values in θ-vector

Figure 5.7: A schematic representation of a θ-vector in BP-volume (a) and dissimilarity of values of the θ-vector (b). Outliers are marked by triangle markers.

to be a non-increasing function

$$\omega = \left(\frac{1-d}{1+\alpha d}\right)^{\beta}, \; \omega \in (0,1). \tag{5.6}$$

This function is called a *correction function*. The parameters α and β control the steepness of the correction curve, see Fig. 5.8a and Fig. 5.8b.

All weighting coefficients are less than or equal to one, i.e. they are damping

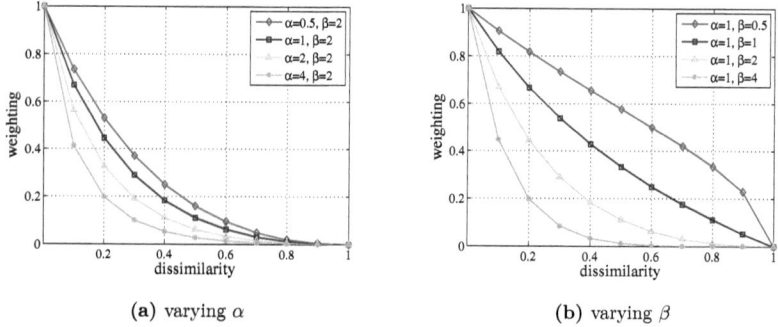

(a) varying α (b) varying β

Figure 5.8: The dependency of the weighting coefficients on the dissimilarity values is called a correction curve. The influence of parameters α (a) and β (b) on the steepness of the correction curve is shown.

coefficients. An amplification of values is not possible. The contribution from all values will be reduced, even from those values, which are smaller than the reference value and potentially do not contribute to the artifacts.

It is important to note that if the relation between the dissimilarity and the weighting coefficients are chosen as a step-function, the weighting scheme is simplified to a smart voting strategy, which allows for seting a flexible threshold for each voxel to control if the projection value is rejected. If M_{xyz} is chosen to be the same value for all voxels and a step-function is used, the algorithm describes the classical voting strategy when projection values larger than a certain absolute threshold are rejected and not used in reconstruction algorithm.

5.3 Non-linear backprojection ωBP for tomosynthesis

The weighting scheme can used within the classical backprojection operator resulting in a novel non-linear backprojection operator. This section is also based on (Levakhina 2012a).

5.3.1 Introducing weighting in the BP operator

The weighting is designed to suppress the contributions from high-absorption features when they potentially contribute to out-of-focus artifacts. Coefficients are calculated individually for each combination of voxel and projection value and are included into the classical backprojection operator, see algorithm 5.1. The weighted version of the backprojection algorithm is denoted by ωBP. The dissimilarity degree is used to identify those voxels for which the given contribution will be "too large" and to assign corresponding weighting coefficients. First, a BP-volume is calculated (line 1 - line 4). Then, for each contribution from each θ-vector a reference value is calculated (line 7) and a dissimilarity degree is assigned (line 8). Based on the dissimilarity degree the weighting coefficients are calculated (line 9) and used within the backprojection operator (line 10).

The BP-volume $h(x, y, z, \theta)$ in tomosynthesis is a four-dimensional space (x, y, z, θ). The direct implementation of the BP-volume calculation for tomosynthesis is infeasible because of size and dimensionality of the data. It is more natural to process the data slice by slice. Additionally, the distance-driven projectors have also a slice-wise nature.

For the slice-wise implementation the following definitions are needed. A *backprojected view* $\mathbb{P}_{\theta,z}(x, y)$ is the result of a simple backprojection operation applied to one projection image $p_\theta(u, v)$, taken along the selected angular view $\theta \in \Theta$, where Θ is a set

Algorithm 5.1: Weighted ωBP (direct implementation)
Input: Projection data $p_\theta(u,v)$, $\theta \in \Theta$
Output: Reconstructed volume $V(x,y,z)$

1 for $\theta \in \Theta$ do // construct BP-volume
2   select $p_\theta(u,v)$;
3   use a stack operator;
4   add the contribution into the BP-volume $h(x,y,z,\theta)$;
5 for $(x,y,z) \in V$ do // introduce weighting in BP
6   construct θ-vector $h_\theta(x,y,z)$;
7   calculate reference value $M_\theta(x,y,z)$;
8   assign dissimilarity degree $d_\theta(x,y,z)$;
9   calculate weighting coefficients $\omega_\theta(x,y,z)$;
10    $V(x,y,z) = V(x,y,z) + \omega_\theta(x,y,z) \cdot \mathbf{BP}\,(p_\theta(u,v))$;

of all projection angles. The backprojection is done on a selected plane of interest z, see equation (5.7). The plane is parallel to the detector plane. A backprojected view is shown in Fig. 5.9a.

$$\mathbb{P}_{\theta,z}(x,y) = \mathbf{BP}_{\theta,z}(p_\theta(u,v)) \qquad (5.7)$$

A *backprojected slice* $\mathbb{S}_z(x,y,\theta)$, or a *backprojected subvolume* is the result of a simple backprojection operation using the given set Θ and the selected coordinate z. The backprojected subvolume $\mathbb{S}_z(x,y)$ is a stack of all backprojected views $\mathbb{P}_{\theta,z}(x,y)$ for the selected coordinate z, see equation (5.8).

$$\mathbb{S}_z(x,y,\theta) = \bigcup_{\forall \theta \in \Theta} \mathbb{P}_{\theta,z}(x,y) \qquad (5.8)$$

A backprojected slice is shown in Fig. 5.9b.

A backprojected volume $\mathbb{V}(x,y,z,\theta)$ is a four-dimensional stack of all backprojected subvolumes $\mathbb{S}_z$, which are located equidistantly for all $z \in V$ see equation (5.9).

$$\mathbb{V}(x,y,z,\theta) = \bigcup_{\forall z \in V} \mathbb{S}_z(x,y,\theta) = \bigcup_{\forall \theta \in \Theta} \bigcup_{\forall z \in V} \mathbb{P}_{\theta,z}(x,y) \qquad (5.9)$$

The BP-volume in tomosynthesis is shown schematically in Fig. 5.9c.

The BP-subvolumes for each slice can be processed separately. Consequently, the dissimilarity and the weighting coefficients can be calculated for each slice location z independently. Weighted backprojection for tomosynthesis can also be done without direct construction of the whole four-dimensional BP-volume at once, see algorithm 5.2.

5.3 Non-linear backprojection ωBP for tomosynthesis

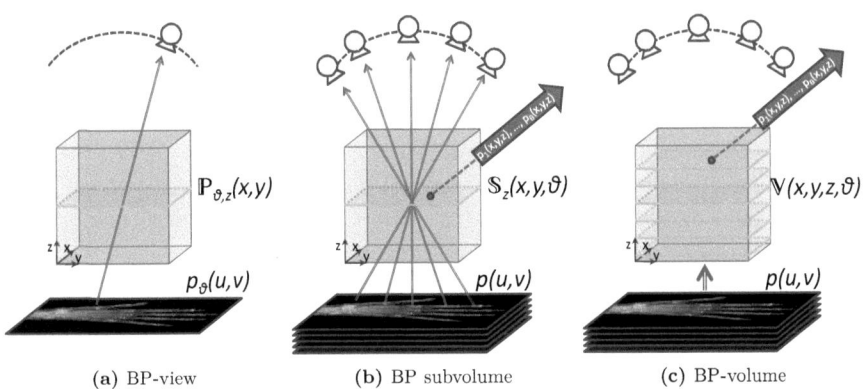

(a) BP-view (b) BP subvolume (c) BP-volume

Figure 5.9: Schematic representation of (a) BP-view; (b) BP-subvolume; (c) BP-volume for tomosynthesis geometry with a fixed detector.

Algorithm 5.2: Weighted ωBP (practically feasible slice-wise implementation)

Input: Projection data $p_\theta(u,v)$, $\theta \in \Theta$
Output: Reconstructed slices $slice_z(x,y)$, $z \in V$

1 **for** $z \in V$ **do** // pre-calculate a set of reference slices
2 $\quad$ calculate $M_z(x,y)$;
3 **for** $\theta \in \Theta$ **do**
4 $\quad$ select $p_\theta(u,v)$;
5 $\quad$ **for** $z \in V$ **do**
6 $\quad\quad$ load reference slice $M_z(x,y)$;
7 $\quad\quad$ caluclate backprojected view $\mathbb{P}_{\theta,z}(x,y)$;
8 $\quad\quad$ assign dissimilarity degree $d_{\theta,z}(x,y,)$;
9 $\quad\quad$ calculate weighting coefficients $\omega_{\theta,z}(x,y)$;
10 $\quad\quad$ $slice_z(x,y) = slice_z(x,y) + \omega_{\theta,z}(x,y) \cdot \mathbf{BP}_z(p_\theta(u,v))$;

For an assignment of a dissimilarity degree for a combination of a z-th $slice_z(x,y)$ and an i-th view $p_{\theta i}(u,v)$ only an i-th BP-view $\mathbb{P}_{\theta i,z}(x,y)$ (line 7) and the reference value $M_z(x,y)$ are required. This way, the weighted backprojection can be easily implemented slice-wise (line 6 - line 10). The reference value can be pre-calculated (line 2).

5.3.2 Demonstration of the dissimilarity and weighting

This subsection demonstrates how the dissimilarity and weighting coefficient images look when the algorithm is applied to real data. Real measured data of a hand has been acquired using the Siemens Mammmomat Inspiration device. The log-transform to convert the intensity images into projection images is done by the device. The hand is approximately 44 mm thick. The reconstructed slices have a spacing of 1 mm. The reference value M is the min value and the correction curve parameters are $\alpha = 1$ and $\beta = 1$.

A dissimilarity and a weighting coefficients image for the slice at 24 mm and an arbitrary view direction at -25° are shown in Fig. 5.10a and Fig. 5.10b. Both, the dissimilarity image and the weighting image are shown with a full [0, 1] window. Bright regions in dissimilarity image denote large dissimilarity degree values. It can be seen, for example, that the region with metacarpal bones has a large dissimilarity (marked by an ellipse). The selected region in the weighting image has relatively small values and is denoted as a dark region, see Fig. 5.10b. Those bones are located outside the selected slice and will contribute to the out-of-focus artifacts, see the unweighted BP-slice in Fig. 5.10c. If the weighted backprojection is done, the contribution of those bones in the current slice is decreased, see Fig. 5.10d.

A dissimilarity and a weighting image for the slice at 35 mm show that in contrast to the slice at 24 mm, the metacarpal bones are in-focus in this slice, see Fig. 5.11a and Fig. 5.11b. Those values contribute with the full contribution to the ωBP. The thumb and the little finger are out-of-focus (marked by ellipses). Correspondingly, the weighting coefficients are small for those regions. The contribution to ωBP of those values is reduced, see Fig. 5.11d.

As a conclusion, the out-of-focus contributions are reduced and the in-plane contributions are preserved when using the proposed weighting algorithm.

5.3 Non-linear backprojection ωBP for tomosynthesis

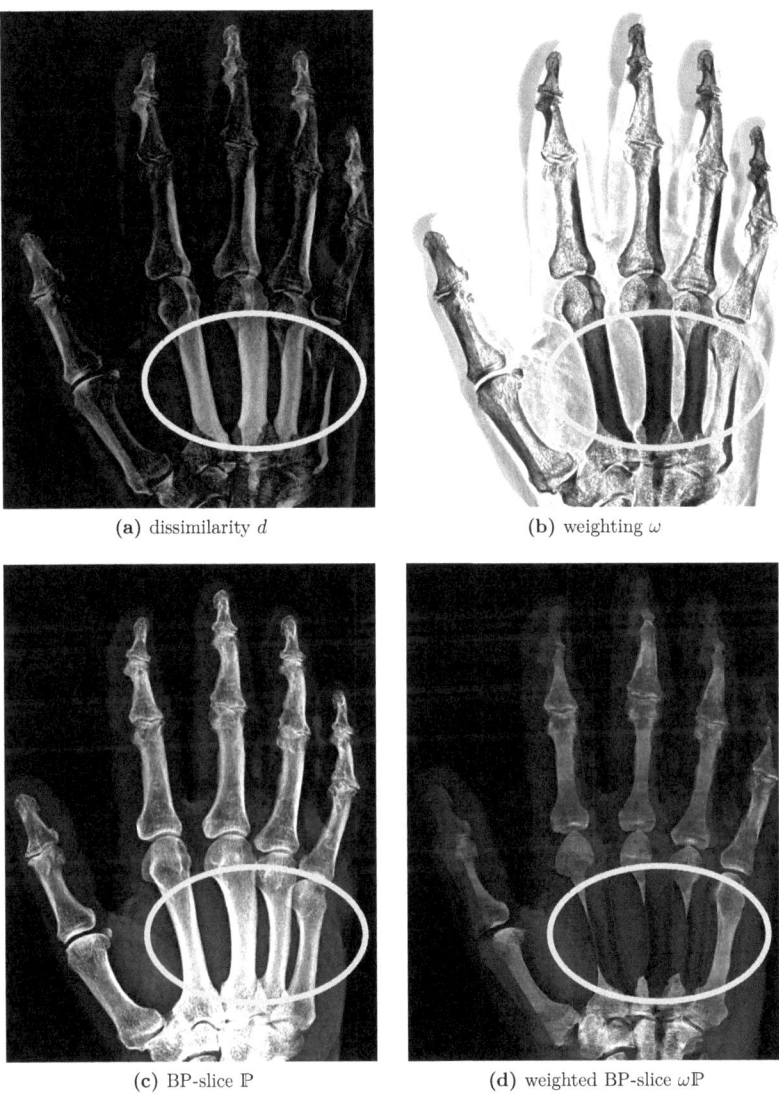

(a) dissimilarity d (b) weighting ω

(c) BP-slice $\mathbb{P}$ (d) weighted BP-slice $\omega\mathbb{P}$

Figure 5.10: Demonstration of dissimilarity and weighting images, slice at 24 mm. The metacarpal bones (marked by an ellipse) are out-of-focus. (a) dissimilarity coefficients, [0 1]; (b) weighting coefficients, [0 1]; (c) an unweighted BP-slice; (d) a weighted BP-slice. It can be seen that out-of-focus contributions from bones are reduced.

Chapter 5. Backprojected space

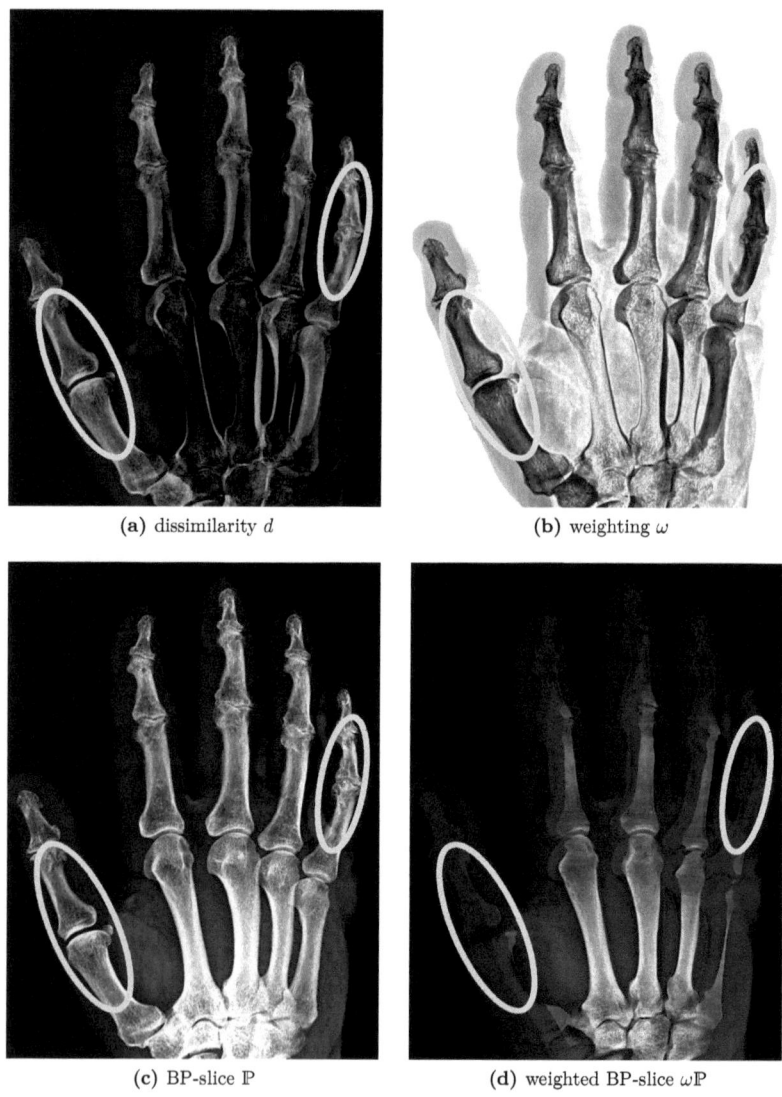

Figure 5.11: Demonstration of dissimilarity and weighting images, slice at 35 mm. The metacarpal bones are in-focus. The thumb and the little finger are out-of-focus. (a) dissimilarity coefficients, [0 1]; (b) weighting coefficients, [0 1]; (c) unweighted BP-slice; (d) weighted BP-slice. It can be seen that out-of-focus contributions are reduced and the in-plane contributions are preserved.

5.3 Non-linear backprojection ωBP for tomosynthesis

5.3.3 Reconstruction results

In this subsection the ωBP reconstruction results with varying parameters of the correction function (M, α, β) will be shown.

The images reconstructed using the ωBP algorithm with the *mean* reference value M are shown in Fig. 5.12a-Fig. 5.12d and the corresponding difference images to the unweighted BP are shown in Fig. 5.12e - Fig. 5.13h. The reconstruction results obtained using the *min* reference value M are shown in Fig. 5.13a-Fig. 5.13d and the corresponding difference images to the unweighted BP are shown in Fig. 5.13e - Fig. 5.13h. The image reconstructed using the unweighted BP is shown in Fig. 5.13a for a visual comparison. All reconstructed images as well as the difference images are comparable with each other, because they are shown using (correspondingly) the same window width and window level.

It can be seen, that the blur is reduced in each image. The steepness of the correction function controls the degree of blur reduction. The usage of the *mean* reference value results in images with fuzzy bone boundaries. The contribution from bone, which is located outside the selected slice is still present, see the left bone in Fig. 5.12. The usage of the *min* reference value results in images with sharp bone boundaries. The contribution from bone, which is located outside the selected slice is partially removed, see Fig. 5.13. At the same time, it is known that the blurred contributions are produced by the out-of-focus structures, which makes it possible to identify them visually. This knowledge is lost when using *min* reference value. It is impossible to identify if the triangle-shaped part of the bone is really in-plane or out-of-plane, but appears sharp because of the chosen reference type.

A too steep correction curve ($\alpha = 2$, $\beta = 2$) results in "over corrected" images. The bone structures are surrounded by undershoot dark shadows for both cases, the *mean* and the *min* reference values, see Fig. 5.12d and Fig. 5.13d. The image reconstructed using the *min* reference value is more noisy because the noise in the projection data is amplified.

As a conclusion, the degree of blur reduction depends on the steepness of the correction function and on the choice of the reference value. The choice of the reference value influences the appearance of the regions, which cannot be unblurred because of incompleteness of the dataset. There is also always a trade-off between the blur reduction and noise amplification.

Chapter 5. Backprojected space

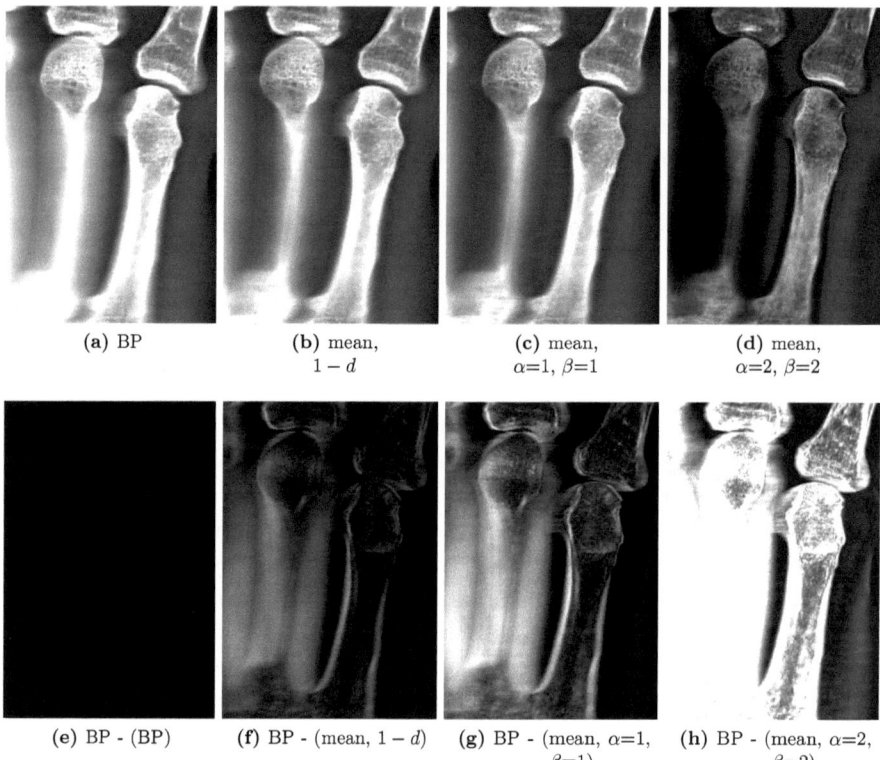

(a) BP (b) mean, $1-d$ (c) mean, $\alpha=1, \beta=1$ (d) mean, $\alpha=2, \beta=2$

(e) BP - (BP) (f) BP - (mean, $1-d$) (g) BP - (mean, $\alpha=1, \beta=1$) (h) BP - (mean, $\alpha=2, \beta=2$)

Figure 5.12: Reconstruction of a hand slice located at 24 mm. (a) unweighted BP; (b)-(d) ωBP with M as the *mean* value and varying α and β; (e)-(h) show the difference between unweighted BP and ωBP with varying wighting parameters.

5.3 Non-linear backprojection ωBP for tomosynthesis

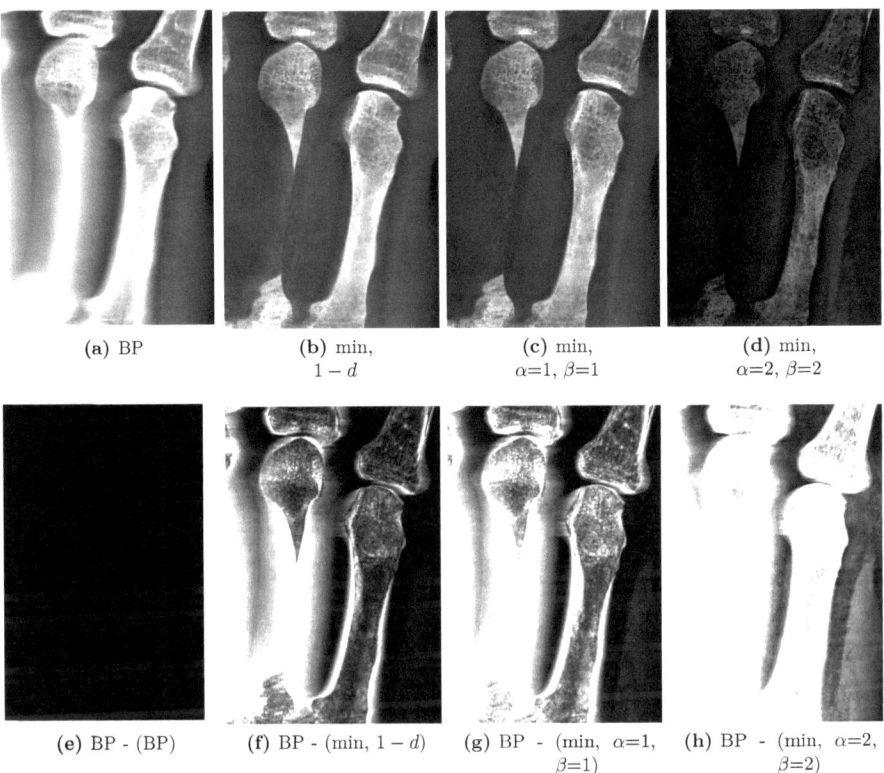

(a) BP (b) min, $1-d$ (c) min, $\alpha=1, \beta=1$ (d) min, $\alpha=2, \beta=2$

(e) BP - (BP) (f) BP - (min, $1-d$) (g) BP - (min, $\alpha=1$, $\beta=1$) (h) BP - (min, $\alpha=2$, $\beta=2$)

Figure 5.13: Reconstruction of a hand slice located at 24 mm. (a) unweighted BP; (b)-(d) ωBP with M as the *minimum* value and varying α and β. (e)-(h) show the difference between unweighted BP and ωBP with varying wighting parameters.

5.4 Weighted ωSART for tomosynthesis

In this section, a non-linear backprojection operator is used within the SART reconstruction algorithm. The ωSART reconstruction results with varying weighting parameters will be shown and compared to the non-weighted SART. More reconstruction results and artifact reduction analysis ca be found in (Levakhina 2013b).

5.4.1 Introducing the non-linear BP into SART

In the presence of high absorption features, such as bones or metal, the corresponding measured projection values will be relatively high. Consequently, the updating term in SART [1] algorithm calculated based on these projection values will be too large for certain voxels, contributing this way to the formation of the artifacts. The non-linear back-distribution of the updating terms can be used to reduce a too large contribution when the updating term is calculated using a large projection value and is back-distributed onto the wrong geometrical location. The weights $\omega_{i,j}$ are used to control the influence of each beam to each voxel in the reconstruction. Weights are calculated using the BP-space representation. The weighting coefficients are calculated based on a dissimilarity measure, which is evaluated in this space. To derive this weighting, the measured projection data is used. This results in a unique set of weighting coefficients for each set of measurements. The SART algorithm with the proposed weighting scheme is denoted as ωSART

$$f_i^{(n+1)} = f_i^{(n)} + \frac{\lambda}{A_{i,+}} \sum_{j \in J_\theta} \omega_{i,j} \frac{A_{i,j}}{A_{+,j}} \left(p_j - \overline{p}_j \left(f^{(n)} \right) \right). \quad (5.10)$$

5.4.2 Computational complexity and implementation strategy

The implementation of ωSART (see algorithm 5.3) is similar to the implementation of the non-weighted SART. The slice-wise implementation strategy of the weight calculations (see line 12 - line 15) can be directly taken from the ωBP algorithm. Weights are produced based on the dissimilarity calculated in the BP-volume. The BP-volume is constructed using the raw-data. The difference is that not the backprojected raw-data is weighted, but the backprojected updating terms (line 19).

---

[1] It is not limited to the algebraic reconstruction and can potentially be included into various iterative reconstruction algorithms with an additive updating strategy. Furthermore, the scheme is not limited to tomosynthesis geometry, but can be extended to the full scan CT case and applied, for example, to the metal artifact reduction problem.

5.4 Weighted ωSART for tomosynthesis

Compared to the unweighted SART algorithm, one iteration of the ωSART algorithm requires an additional computation of the dissimilarity and the weighting coefficients. This is equivalent to $N_\gamma \times N_{slices}$ of U_{BP} units (line 12). It results in total of

$$2 \cdot N_\gamma \times N_{slices} \times (U_{FP} + U_{BP}) + N_\gamma \times N_{slices} \times (U_{BP}) \qquad (5.11)$$

computational costs for one iteration of ωSART. This means that one iteration of ωSART is $N_\gamma \times N_{slices} \times U_{BP}$ more expensive[2] than one iteration of the non-weighted SART. The reference slices M can be pre-computed (line 4). It requires additional $N_x \times N_y \times N_{slices}$ units of memory.

Algorithm 5.3: Slice-wise implementation of ωSART for tomosynthesis

Input: Projection data $sino(\theta)$, $\theta \in \Theta$
Output: Reconstructed volume $slice(z)$, $z \in V$

1 **Initialization** : dictionaries for $sino\_name$, $sino\_upd\_name$ $\forall \theta \in \Theta$;
2 dictionaries for $slice\_name$, upd_term_name $\forall z \in V$;
3 $slice$ $\forall z \in V$;
4 dictionary for M_name $\forall z \in V$;
5 pre-compute reference slices $M(z)$ $\forall z \in V$;
6 **for** $\theta \in \Theta$ **do**
7 $\quad$ read $sino\_name$ and $sino\_upd\_name$ for given θ from dictionary;
8 $\quad$ $sino\_upd(\theta) = \mathbf{FP}\left(slice\left(\forall z\right), \theta\right)$;
9 $\quad$ $A_{+,\theta} = \mathbf{FP}\left(slice\_ones\left(\forall z\right), \theta\right)$;
10 $\quad$ $sino\_upd(\theta) = (sino\_upd(\theta) - sino(\theta))/A_{+,\theta}$;
11 $\quad$ **for** $z \in V$ **do**
12 $\quad\quad$ $\mathbb{P}(\theta, z) = \mathbf{BP}_z\left(sino(\theta)\right)$;
13 $\quad\quad$ read M_name for given z from dictionary;
14 $\quad\quad$ $d(z) = (abs(\mathbb{P}(\theta, z) - M(z))/range)$;
15 $\quad\quad$ $\omega(z) = \left(\frac{1}{1+\alpha d(z)} \frac{d(z)}{}\right)^\beta$;
16 $\quad\quad$ read upd_term_name and $slice\_name$ for given z from dictionary;
17 $\quad\quad$ $upd\_term(z) = \mathbf{BP}(sino\_upd(\theta), z)$;
18 $\quad\quad$ $A_{z,+} = \mathbf{BP}(sino\_ones(\theta), z)$;
19 $\quad\quad$ $slice(z) = slice(z) - (\lambda \cdot \omega(z) \cdot upd\_term(z))/A_{z,+}$;
20 $\quad\quad$ clear $upd\_term(z)$, $A_{z,+}$;
21 $\quad$ clear $sino\_upd(\theta)$, $A_{+,\theta}$;

---

[2] The U_{FP} and U_{BP} computational units have been introduced and described in chapter 4.

5.4.3 Reconstruction results

Real measured data of a hand have been used for the reconstruction. The measured data have been acquired using the Siemens Mammmomat Inspiration device. The intensity images have been transformed into projection images using a log-transform and an additional empty scan I_0. The following reconstruction parameters have been used for both, SART and ωSART algorithms: the iteration process has been stopped after 3 iterations, the relaxation parameter λ is 0.3, the initial guess is zero, the projection access order is a random permutation and no roughness prior functions have been involved. Practice has shown that those parameters result in acceptable images with short computation time. For each object a stack of slices with 1 mm thickness has been reconstructed. All images in the stack are parallel to the detector plane. The reconstructed volume of the hand contains 44 slices. Window width and window level have been adjusted in order to emphasize artifacts. The tube movement direction is from the left to the right with respect to all presented tomosynthesis images. The reconstructed images of the hand demonstrate artifact reduction in a clinical case.

The ωSART reconstruction results of the hand obtained using varying reference value M and the parameters of the correction function are shown in Fig. 5.14b - Fig. 5.14g. The shape of the correction function varies from a linear ($\alpha = 0, \beta = 1$) to a non-linear steep curve ($\alpha = 2, \beta = 2$). The image, reconstructed using classical SART, is shown for comparison in Fig. 5.14a. Out-of-focus artifacts appear as ghosting copies of the artifact-causing features. High-absorption features at the in-plane slice are surrounded by dark shadows. Artifacts are marked by white arrows. Slices reconstructed using the ωSART method have less out-of-focus artifacts and they are less affected by the dark shadows. The degree of artifacts reduction depends on the steepness of the correction curve, which is similar to the simple ωBP. The reference value $M = mean$ results in better artifacts reduction compared to $M = min$ with fixed α and β parameters. However, $M = min$ is also responsible for producing additional artifacts, see the region of interest marked by an ellipse. In this region, the out-of-focus artifacts produced by metacarpal bones are reduced. The boundaries of metacarpal bones are sharp, although they do not belong to the shown slice.

The axial slices through the hand at $y = 800$ and $y = 1700$ are shown in Fig. 5.15. The ωSART reconstruction ($M = min$, $\alpha = 2$, $\beta = 2$) shows great reduction of the limited angle artifacts also in z-direction. The streak-like artifacts and dark shadows produced by bones (marked by white arrows) are reduced. However, the shape of features in z-direction cannot be completely recovered and the triangle-shaped distortion of the bones is still present.

5.4 Weighted ωSART for tomosynthesis

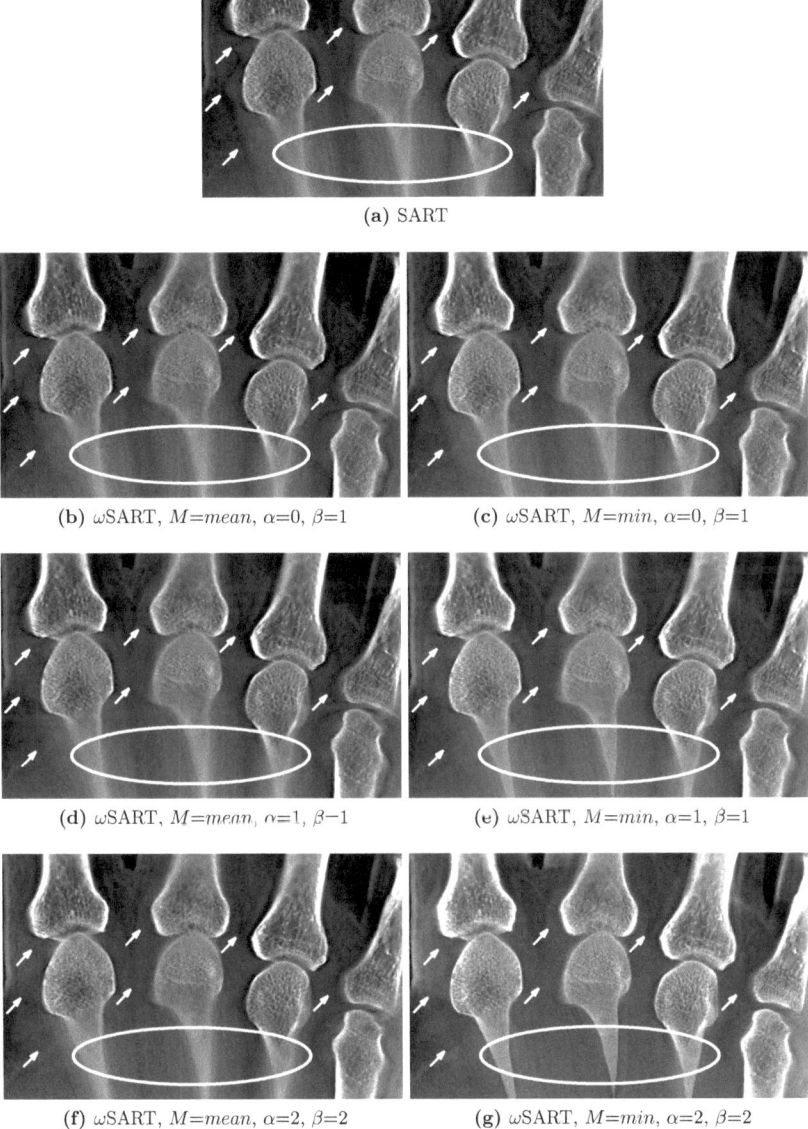

(a) SART

(b) ωSART, $M=mean$, $\alpha=0$, $\beta=1$

(c) ωSART, $M=min$, $\alpha=0$, $\beta=1$

(d) ωSART, $M=mean$, $\alpha=1$, $\beta=1$

(e) ωSART, $M=min$, $\alpha=1$, $\beta=1$

(f) ωSART, $M=mean$, $\alpha=2$, $\beta=2$

(g) ωSART, $M=min$, $\alpha=2$, $\beta=2$

Figure 5.14: SART and ωSART reconstruction of the hand. The shape of the correction function varies from a linear ($\alpha = 0, \beta = 1$) to a relatively steep curve ($\alpha = 2, \beta = 2$). White arrows point to the limited angle artifacts which are reduced in ωSART reconstruction. The white ellipse points to artifacts which are highlighted when $M = min$ is used.

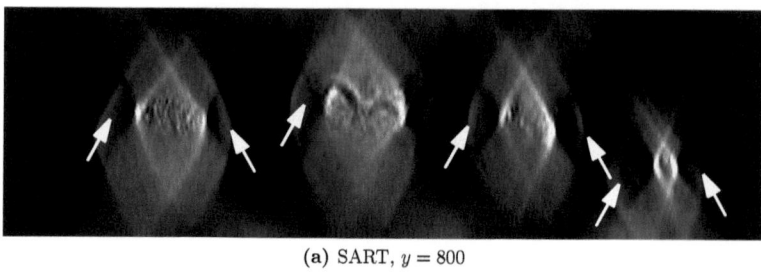

(a) SART, $y = 800$

(b) ωSART, $y = 800$, $M=min$, $\alpha=2$, $\beta=2$

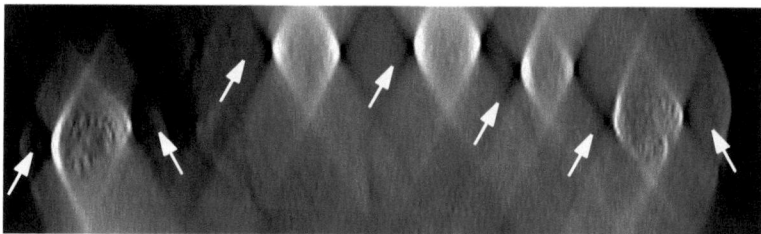

(c) SART, $y = 1700$

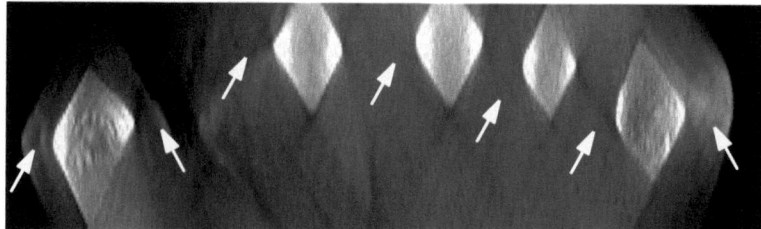

(d) ωSART, $y = 1700$, $M=min$, $\alpha=2$, $\beta=2$

Figure 5.15: SART and ωSART ($M = min, \alpha = 2, \beta = 2$) reconstruction of the hand. Axial cuts located at at $y = 800$ and $y = 1700$ are shown. White arrows point to the limited angle artifacts which are reduced in ωSART reconstruction.

5.5 Weighted ωSART for metal artifact reduction in CT

The ωSART algorithm can be extended for the full-angle 360° data and can be used for the metal artifact reduction problem to correct for data inconsistency inside the metal trace. The data inconsistency causes metal artifacts in the reconstructed images. Usually, the metal-related data-samples are considered as invalid and are replaced by the artificial data, produced by an interpolation (inpainting). Alternatively, a reconstruction algorithm can be modified to ignore or to weight the inconsistent data, see e.g. (Oehler 2007). For both types of approaches the metal segmentation step is required in order to identify the metal-related projections. If the metal segmentation is done in the image domain, a preliminary reconstructed image is needed. The accuracy of the segmentation is influenced by the metal artifacts. The presence of a large amount of metal makes it almost impossible to segment the metal accurately because of severe artifacts. The segmentation in the sinogram domain cannot be done using simple thresholding. It is time-consuming if it is done manually or computationally expensive if it is done using dedicated registration algorithms. The weighting scheme presented in the previous sections can be used to control the contribution of metal projections without preliminary segmentation of the metal. The metal is preserved on its correct location.

Some preliminary results of the ωSART reconstruction applied to the metal artifact reduction problem will be shown below. The non-weighted SART reconstruction results will be shown for the comparison. The reference value M is min for all shown results. The projection data of a torso phantom with three metal rods and the projection data of a patient with a hip implant have been acquired using the Siemens Somatom Definition AS device.

5.5.1 Parameter γ

In tomosynthesis geometry the intersection length through the object is almost the same for each pixel. Therefore, the variations in the projection data are caused only by the material density. In case of clinical CT, the object dimensions might vary from different sides, which means that the different rays will have different length through the object. The projection values are influenced not only by density but also by the the intersection length. Even in an ideal case of a homogeneous object, the measured values might differ from each other, although rays travel through the same type os tissue. Therefore, a third parameter for the correction curve is needed to control what degree of dissimilarity is still allowed without a weighting

$$\omega = \left(\frac{1-d}{\gamma + \alpha d}\right)^{\beta}. \tag{5.12}$$

The parameter γ is a new parameter compared to the tomosynthesis case. It controls the weighting of the data with small dissimilarity degree (i.e. similar data). The smaller γ, the larger the dissimilarity value at which the correction curve starts to drop down from the value $\omega=1$, see Fig. 5.16. The smaller γ, the larger amount of data will be weighted with weighting $\omega \geq 1$;

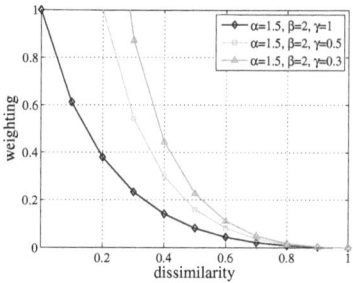

Figure 5.16: Influence of γ on the behavior of the correction curve $\omega(d)$

Another difference is that the weighting coefficients are allowed to be larger than 1. It is important to check that the weighting does not result in an explosion of the iteration process. Typically, the relaxation parameter range $0 < \lambda \leq 2$ is recommended for non-weighted SART, therefore we assume that $0 < \lambda \times \omega \leq 2$ is a reasonable limitation.

The histogram of the dissimilarity shows the data distribution visually, see Fig. 5.17. The parameter *range* stretches the histogram. The smaller the parameter *range* the wider the histogram. The *range* = 4.6 is a range of projections data without metal. The *range* = 7.6 includes some portion of metal data. The correction curve with fixed parameters results in different distributions of weighting coefficients for different *range* values. The distribution of weighting coefficients is re-distributed closer to zero for *range* = 4.6. It means that more contributions got smaller weighting coefficients.

5.5 Weighted ωSART for metal artifact reduction in CT

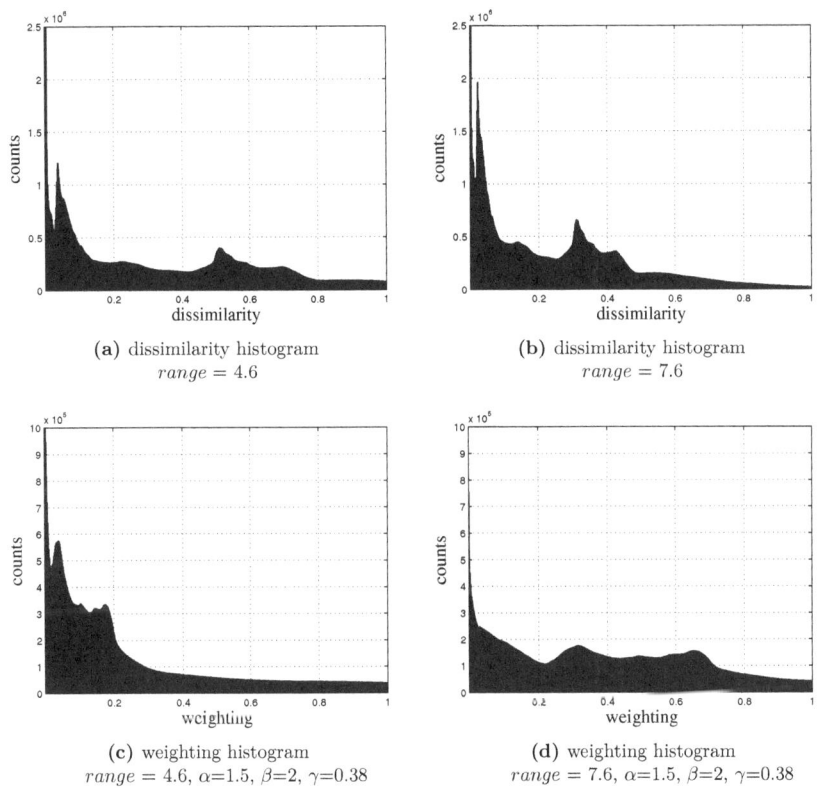

(a) dissimilarity histogram $range = 4.6$

(b) dissimilarity histogram $range = 7.6$

(c) weighting histogram $range = 4.6$, $\alpha=1.5$, $\beta=2$, $\gamma=0.38$

(d) weighting histogram $range = 7.6$, $\alpha=1.5$, $\beta=2$, $\gamma=0.38$

Figure 5.17: The parameter $range$ stretches a histogram of dissimilarity. The $range = 4.6$ is a range of projections data without metal. The $range = 7.6$ includes some portion of metal data. The same parameters of correction curve results in different distribution of weighting coefficients for different $range$ values. For $range = 4.6$, more weighting coefficients are concentrated close to zero.

5.5.2 Reconstruction results

The results of the first sub-iterations of the iterative scheme for the phantom with three metal rods are shown in Fig. 5.18. The number of used projections are 5, 15 and 65. The SART images (Fig. 5.18a - Fig. 5.18c) show the homogeneous distribution of the high-intensity metal-based updating term along the ray-lines. It potentially contributes to the streak-like noise and metal artifacts. The weighted ωSART images show that the contribution of the metal-based projections are reduced when those values are backprojected outside the metal region.

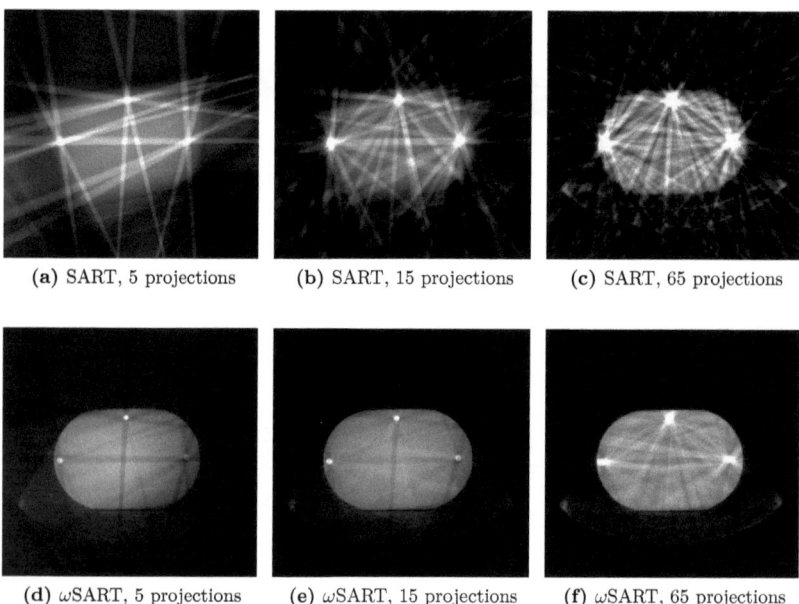

(a) SART, 5 projections   (b) SART, 15 projections   (c) SART, 65 projections

(d) ωSART, 5 projections   (e) ωSART, 15 projections   (f) ωSART, 65 projections

Figure 5.18: Visual demonstration of the weighting scheme. The results of the SART and ωSART iterative scheme using the first projections are shown.

The reconstruction results of SART and ωSART of the phantom with three metal rods are in Fig. 5.19. The iteration process is stopped after 10 iterations. A non-negativity constraint and a relaxation parameter $\lambda = 0.15$ were used. No smoothing priors or additional data post- and pre-processing steps were included. The image reconstructed using the unweighted SART reconstruction (Fig. 5.19a) shows typical metal artifacts and streak-like noise. The ωSART (Fig. 5.19b - Fig. 5.19d) reduces

5.5 Weighted ωSART for metal artifact reduction in CT

metal artifacts and streaks. The images show that the degree of artifact reduction depends on the chosen parameters. Almost no additional artifacts are introduced in the reconstructed images. The optimal parameters for the weighting curve were found empirically. Optimal parameters for the phantom are $\lambda=0.15$, $\alpha=1.5$, $\beta=2$, $\gamma=0.38$, $range=17$.

As an additional example, reconstructing results of a patient with a hip implant are shown in Fig. 5.20. Two slices with different amount of metal have been chosen. The images reconstructed using the unweighted SART algorithm also contains metal artifacts and streak-like noise (Fig. 5.20a). The amount of artifacts and streaks depends on the amount of metal. Metal artifacts hide an anatomy between and around implants. In both cases the ωSART reconstruction (Fig. 5.20b, Fig. 5.20d) shows good ability to suppress streak-like noise. The anatomy in the regions left, right and between the metals, where dark shadows appeared in the unweighted SART reconstruction, is recovered. This it can be seen especially in the second slice with the large amount of metal. At the same time, some additional streak-like artifacts are introduced in the reconstructed images. Those streaks have less intensity than metal artifacts and do not hide anatomy. Nevertheless, they can be seen and corrupt the appearance of the images.

A comparison of the original sinogram of the hip and the simulated sinograms based on the SART and ωSART reconstruction is shown in Fig. 5.21. The original sinogram of the hip with metal implant, two simulated sinograms based on the SART and ωSART reconstruction and their difference to the original sinogram are shown. The region of visible data inconsistency, where two metal traces are crossed, is marked by an ellipse. Due to the beam hardening effect, the measured sinogram does not represent a sum of sinusoidal curves in this region as it is required by the Radon transform. The SART reconstruction results in an object which "fits" into the inconsistent sinogram and therefore has an artificial dark shadow strike between the metal implants. It can be seen, that the ωSART algorithm is able to recover the intensity of the sum of sinusoidal traces in this region, which results in less artifacts in the reconstructed images and large values in the difference image in those regions. The overall difference excluding the inconsistent regions is smaller for the ωSART simulated sinogram [1]. The ωSART is able to recover the intensity of sinusoidal traces in this region.

Although the ωSART reconstruction shows potential to reduce metal artifacts successfully, it needs further investigation and optimization. The question how to select optimal parameters is still open. Additionally, a comparison to classical metal artifact reduction techniques (i.e. a linear interpolation) is required.

---

[1] The horizontal lines in the difference images are caused by the horizontal lines in the original data. It might happen due to the X-ray current fluctuation in the acquisition.

Chapter 5. Backprojected space

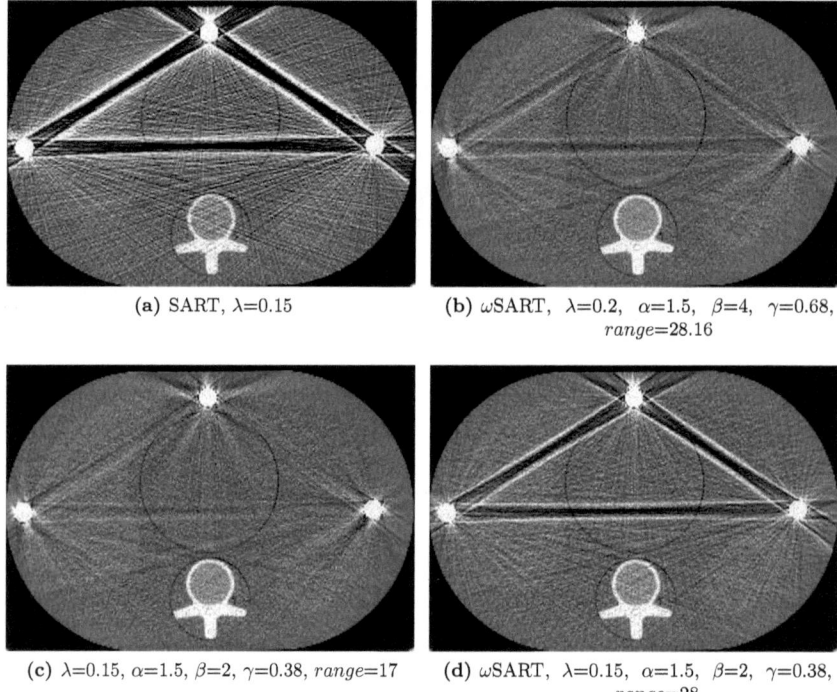

(a) SART, λ=0.15

(b) ωSART, λ=0.2, α=1.5, β=4, γ=0.68, $range$=28.16

(c) λ=0.15, α=1.5, β=2, γ=0.38, $range$=17

(d) ωSART, λ=0.15, α=1.5, β=2, γ=0.38, $range$=28

Figure 5.19: The unweighted SART reconstruction of a phantom with three metal rods (a) shows typical metal artifacts and streak-like noise. The novel ωSART (b-d) reduces the metal artifacts. The degree of artifact reduction depends on chosen parameters. No additional data post- and pre-processing steps were included in ωSART algorithm.

5.5 Weighted ωSART for metal artifact reduction in CT

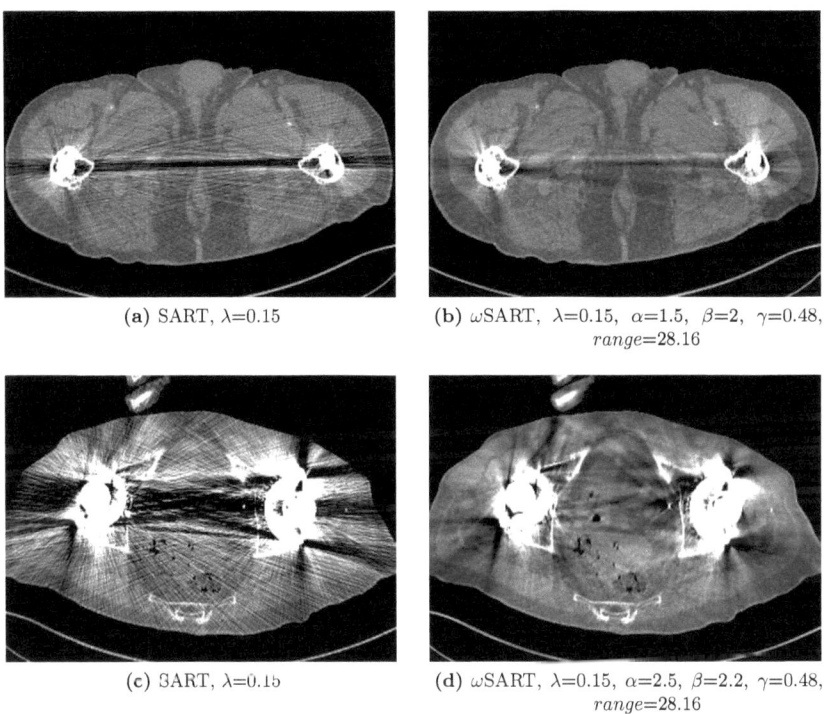

(a) SART, $\lambda=0.15$

(b) ωSART, $\lambda=0.15$, $\alpha=1.5$, $\beta=2$, $\gamma=0.48$, $range=28.16$

(c) SART, $\lambda=0.15$

(d) ωSART, $\lambda=0.15$, $\alpha=2.5$, $\beta=2.2$, $\gamma=0.48$, $range=28.16$

Figure 5.20: The unweighted SART reconstruction of a hip with metal implants (a), (c) shows typical metal artifacts and streak-like noise. The ωSART (b),(d) reduces metal artifacts. The degree of artifact reduction depends on chosen parameters. No additional data post- and pre-processing steps were included in the ωSART algorithm.

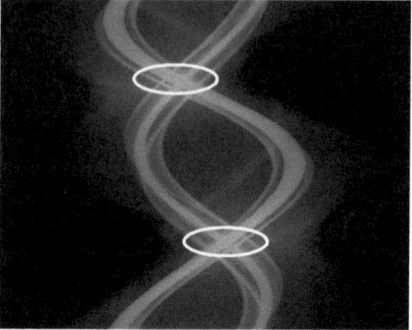
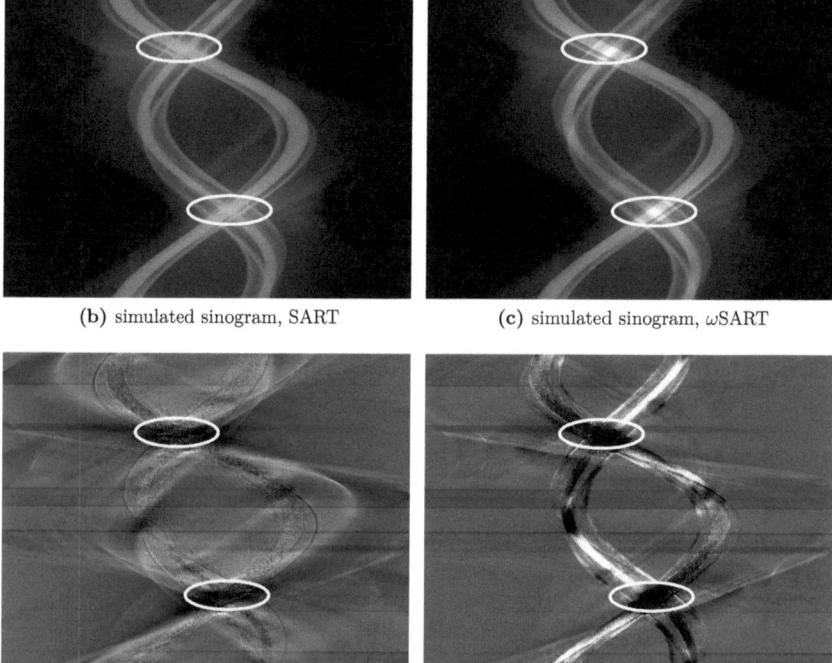

Figure 5.21: Comparison of sinograms. (a) original sinogram of a hip with metal implant; (b-c) simulated sinogram of SART and ωSART reconstruction; (c-d) difference to the original sinogram. The region of visible data inconsistency, where two metal traces are crossed, is marked by an ellipse. ωSART is able to recover the intensity of sinusoidal traces in this region.

5.6 Interpolation in BP-space for metal artifact reduction in CT

As it has been mentioned in the previous section, the metal influenced data in the sinogram can be considered as invalid and therefore can be removed from the dataset. The resulting gap has to be filled with some artificial data. Interpolation for producing artificial data instead of invalid metal data can be done not only in the sinogram domain but also in the BP-space. The properties of BP-space offer new flexible opportunities for data interpolation along the sinusoidal traces

5.6.1 θ-interpolation in BP-space

Lets consider a sinogram $p(\xi, \theta)$, $\theta \in [0, 2\pi)$. Consider that $p(\xi, \theta)$ contains one or more sinusiod-like curves influenced by the metal object. Those metal traces can be detected either by manual segmentation or by using the registration algorithms in sinogram domain. Alternatively it can be done by creating a binary metal mask using a preliminary reconstruction. Regardless of the detection method, after removing the data from the sinogram, it will contain a gap with no data. The sinogram with the gap can be transformed in the BP-space. Each xy-plane will contain some missing ridge lines, produced by the gap, see Fig. 5.22a.

One dimensional interpolation in the xy-plane perpendicular to ridge lines is equivalent to one-dimensional view-wise interpolation in ξ direction in the sinogram domain. It can be performed fast and easily in sinogram domain and there is no need to construct a BP-space in this case.

The advantage of BP-space representation for the interpolation purposes can be seen when looking in θ-direction. As it has been presented in section 5.1, moving along the ridge line and subsequently in θ-direction is equivalent to moving along all lines within a peanut-shape. Similarly, gaps with missing data, which are parallel to the ridge function in each xy-plane, can be used to navigate through the sinusoidal curves in each peanut corresponding to each missing point. In other words, interpolation in θ-direction is equivalent to interpolation along each sinusoidal curve which crosses the selected point (ξ', θ'), i.e. a set of curves within a peanut-shape of (ξ', θ').

For the interpolation a BP-volume of the data with a gap and a BP-volume of the binary mask have to be constructed, see algorithm 5.4. Then, for each coordinate (x, y) a corresponding θ-vector has to be selected. At some coordinate positions, the θ-vector might contain missing values in the first $n_{missing}$ or the last $n_{missing}$ entries. It means that no interpolation is possible because there is no data available on one side. The periodicity property of sinogram can be used to enable the interpolation in those cases. The data in the θ-vector can be rearranged periodically in a way, that the gap (each gap) will be surrounded by valid data, see algorithm 5.5.

Algorithm 5.4: Interpolation in BP-volume in θ-direction

Input: Projection data with a gap $p(\xi, \theta)$, $\theta \in \Theta$
binary mask of a gap $mask(\xi, \theta)$, $\theta \in \Theta$
Output: BP-volume with interpolated data in the gap $h_{new\,data}(x, y, \theta)$, $\theta \in \Theta$
1 create a BP-volume of the data with gap $h(x, y, \theta) = \$(p(\xi, \theta))$;
2 create a BP-volume of the binary mask $m(x, y, \theta) = \$(mask(\xi, \theta))$;
3 **for** $(x, y) \in V$ **do**
4   select a data θ-vector $h_{xy}(\theta)$;
5   select a mask θ-vector $m_{xy}(\theta)$;
6   rearrange $h_{xy}(\theta)$ periodically according to the gap $m_{xy}(\theta)$;
7   do interpolation of $h_{xy}(\theta)$ in the gap;
8   write results in $h_{new\,data}(x, y, \theta)$;

Algorithm 5.5: Periodic rearrangement of $h_{xy}(\theta)$ according to the longest subsequence in the gap $m_{xy}(\theta)$

Input: θ-vector of the data h
θ-vector of the mask m
Output: periodically re-arranged θ-vector h'
1 find the longest subsequence $m_{longest}$ in the mask m;
2 find the index of the middle element i_{mid} of $m_{longest}$;
3 assign idx_{mid}-th element of h as the first element: $h'(1) = h(idx_{mid})$;
4 **for** $i \in [1, N_\theta - 1]$ **do**
5     $idx_{new} = idx_{mid} + i$;
6   **if** $idx_{new} > N_\theta$ **then** // index is outside the range
7        $idx_{new} = idx_{new} - N_\theta$;
8     $h'(i+1) = h(idx_{new})$;

5.6.2 Preliminary results

For the simulation study of the θ-interpolation in BP-space the standard Shepp-Logan phantom was used. It has a size of 300×300 pixels with a pixel size of 0.1 cm. The missing data is arbitrary located at the lower left corner of the phantom at ($x_g = 108$, $y_g = 233$)) with the diameter of 1 cm (10 pixels).

The interpolation in θ-direction (algorithm 5.4) with data periodicity rearrangement (algorithm 5.5) and a linear interpolation method has been performed in the gap. The data with gap and interpolation results are shown in Fig. 5.22. An xy-plane (Fig. 5.22b) shows that the interpolation works well in regions which are far away from the gap. The interpolation in regions closer to the gap results in data which looks not fitting to the rest of the data. The data in the gap is completely missed. It can be seen in Fig. 5.22c and Fig. 5.22d that on the position of the gap there is no data available in θ-direction for interpolation. Additionally, close to the region of the gap, interpolation does not work competely. It results in vertical lines along the gap. The distance between points in θ-direction is too large and interpolation is done between points in sinogram which are located too far away from each other. At the same time, when the gap trace in BP-space is parallel to the θ-axis, interpolation provides good results, see Fig. 5.22e and Fig. 5.22f.

Obviously, the rest of the gap and the not-well fitting data will cause artifacts close to the position of the "metal" which has been removed. At the same time, regions which are far away from the metal position are filled with data which visually fits the rest of the data. We can expect that in those regions no new artifacts will be introduced.

For further work, the interpolation in θ-direction can be combined with in-plane interpolation in order fill the data points which are completely missed. Additionally, the reconstruction strategies should be considered. The BP-space can be transformed back to sinogram domain and reconstructed with any of the conventional reconstruction algorithms. Hoverer, the averaging process in the inverse stack operator might result in averaging of interpolation error and artifacts in the reconstructed images. Alternatively, the reconstruction can be done directly in the BP-space. For this, the reconstruction algorithms need to be adapted to usage in the BP-space or new algorithms need to be developed.

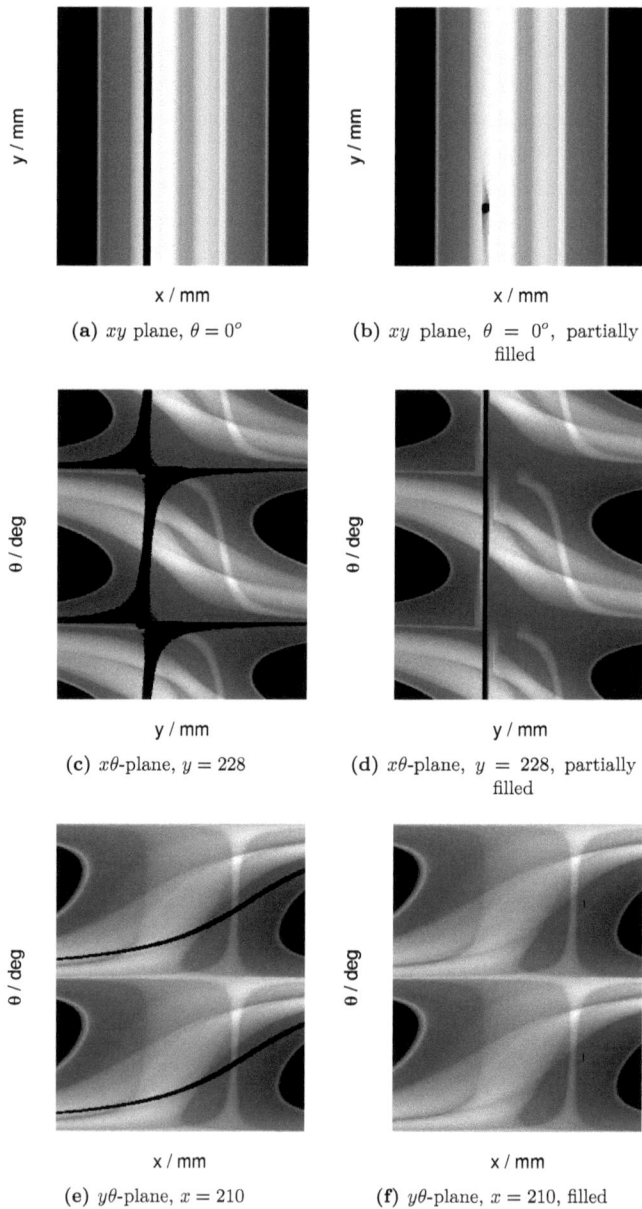

Figure 5.22: Illustration of the θ-interpolation in BP-space. (a), (c), (e) show planes with a gap in BP-volume; (b), (d), (f) show the interpolated planes in BP-volume.

Chapter 6

Dual-axis tilt acquisition geometry for musculoskeletal tomosynthesis

Contents

6.1	Tomosynthesis "mini" simulator	138
6.2	Influence of the system acquisition parameters	142
6.3	A novel geometry: hybrid dual-axis tilt acquisition	151
6.4	Influence of the object orientation	161

This chapter is based on (Levakhina 2012b, Levakhina 2013a) and presents a novel acquisition geometry for tomosynthesis imaging of hands.

The characteristics of the reconstructed images in tomosynthesis depend on several factors. In chapter 3 it has been discussed that the practical implementation strategy influences the accuracy of the reconstruction. The choice of the reconstruction algorithm and its parameters, as it has been addressed in chapter 4 and chapter 5, influences the appearance of the artifacts. In this chapter, the influence of several imaging system parameters and the acquisition geometry on the tomosynthesis performance will be discussed. The main focus of this chapter is to present a novel hybrid dual-axis tilt acquisition geometry, which is superior to the standard acquisition geometry. In the first part of this chapter, the potential and limitations of the standard geometry will be discussed. The comprehensive understanding of the effect of system acquisition parameters on the reconstructed image quality is very important, therefore, it includes a literature review as well as a simulation study. In the second part of this chapter, the alternative geometry will be presented. Its theoretical background will be discussed

138   Chapter 6. Dual-axis tilt acquisition geometry

and software simulation results will be shown. Additionally, the effect of the object orientation will be discussed and simulated. Results will be presented for both, the standard and the novel dual-axis geometry. Finally, suggestions for the future work will be given.

6.1 Tomosynthesis "mini" simulator

Most of the tomosynthesis performance studies, which can be found in literature, are based on characteristics, which can be easily described mathematically. They are typically measured using special phantoms rather than anthropomorphic phantoms. For example, the resolution and the artifacts can be measured using a sharp edge and described using the SNR, MTF, NPS, PSF and ASF. It is not straight-forward to apply those metrics to imaging of clinically relevant objects. Sharp edges for MTF calculation or large homogeneous regions for SNR calculation are not guaranteed to be present in such objects. Typically, the reconstructed images of anthropomorphic phantoms and real patients are evaluated only visually or based on observer studies. A simulation software and digital phantoms allow for measuring image quality based on full-reference metrics. It can be represented as one quantitative parameter, which is easy to interpret. Clinically relevant objects can be used rather than simple geometrical phantoms. The whole reconstructed volume can be used at once instead of a ROI or a specific feature (e.g. an edge).

6.1.1 A finger bone software phantom

A high-resolution three-dimensional volume of a dried human finger bone has been used as a software phantom. In order to create this phantom, the bone was measured using the Skyscan1172 micro-CT and the reconstruction was done by the device using the build-in Feldkamp reconstruction algorithm (FDK) (Feldkamp 1984). The reconstructed volume has been post-processed using denoising and background removal. Slices in z-direction have been averaged for the slice thickness of 0.2 mm. The resulting 30 slices have been used for the simulation of projections and as a reference for the quantitative evaluation of the reconstructed results.

Three orthogonal slices of the bone phantom are shown in the Fig. 6.1. The bone phantom represents an anatomy of fine trabecular structures, which are of the interest for tomosynthesis imaging of hands. The selected ROIs are marked by ellipses. The ROI1 is located in an in-plane region without the tissue to show later the formation of

6.1 Tomosynthesis "mini" simulator

the out-of-focus artifacts. The ROI2, ROI3 and ROI4 represent a bone boundary in the in-plane slice and two orthogonal slices, respectively.

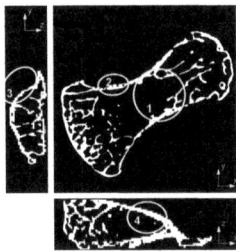

Figure 6.1: A bone software phantom based on the micro CT FDK reconstruction. Three orthogonal slices are shown. Four regions of interests are marked with ellipses. An orientation angle with respect to the X-ray tube rotation axis is 45°

For the study of the influence of the object orientation, the bone phantom can be rotated in the xy-plane. The object orientation angle is defined as an angle between the longest dimension of the object and the y-axis. In Fig. 6.1 the orientation angle of the bone is 45°.

6.1.2 Simulation software

For this study, a software simulation framework (MATLAB/C++) for tomosynthesis acquisition and reconstruction has been developed. The Siemens Mammomat Inspiration device geometry has been used. The device is equipped with a large flat-panel detector and a half-cone X-ray tube. The angular range of this device is 50° and the number of the acquired projections is 25. The typical size of a hand is 220 x 130 mm² and the thickness is 30-50 mm. The size of the bone phantom is approximately 20 x 10 mm² and the thickness is 6 mm. Therefore, the geometry of the device has been scaled down to match physical dimensions of the detector and the used phantom. The scaling factor equals five, see Fig. 6.2 and compare to Fig. 2.8. The orientation of the reconstructed tomosynthesis slices is parallel to the detector plane. The reconstructed slice thickness is 0.2 mm and the pixel element size is 0.017 mm. The imaging and reconstruction parameters are summarized in the Table 6.1.

The main goal of the simulations was to investigate the tomosynthesis performance in dependence of acquisition parameters. An ideal system was considered and no physical limitations were taken into account. The X-ray source is a point source and the spectrum is monoenergetic. Noiseless projections have been simulated. The forward- and

Chapter 6. Dual-axis tilt acquisition geometry

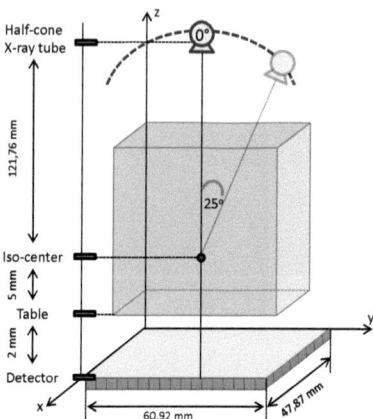

Figure 6.2: Tomosynthesis "mini"-geometry. This geometry was used for the simulation of projections and image reconstruction in the performance evaluation study.

Table 6.1: Tomosynthesis geometry and imaging parameters.

	System parameter	Value
Distances	Source-to-Iso	121.7 mm
	Iso-to-Table	8.2 mm
	Table-to-Detector	4.8 mm
Detector	Detector length x	60.9 mm
	Detector length y	47.8 mm
	Dexel size	0.017 mm
Reco	Slice size	3584 x 2816
	Pixel size	0.017 mm
	Slice thickness	0.2 mm

backprojections have been modeled using the distance-driven method (De Man 2004) adapted for the fixed-detector tomosynthesis geometry (Levakhina 2011a).

The simulator works in two modes: the standard acquisition mode and the dual-axis tilt acquisition mode. The geometry with fixed detector, lying in the xy-plane and the X-ray tube which moves along a one-dimensional arc around the x-axis above the detector is referred to as the standard acquisition mode. The geometry, which acquires data outside the one-dimensional arc, is referred to as the dual-axis tilt acquisition mode. The detailed description of the dual-axis mode will be given in the section 6.3.

6.1 Tomosynthesis "mini" simulator

The simulation of the dual-axis tilt acquisition geometry results in data acquired over a non-standard acquisition trajectory. This data has to be reconstructed. The well-known FBP reconstruction cannot be used in this study, because two different filter kernels are required for two different geometries. The design of the filter influences the image quality, which make the reconstruction results incomparable. The SART algorithm was used in this study, because it can be applied to both geometries without major modifications.

6.1.3 Image quality metrics for performance evaluation

The tomosynthesis performance in dependency of the imaging parameters and geometry has been evaluated quantitatively and qualitatively.

6.1.3.1 Quantitative reference-based evaluation

The reconstructed image have been quantitatively compared using the well-known Normalized Root Mean Squared Error (NRMSE). The NRMSE has been used to measure the similarity degree between the software phantom volume and obtained reconstructed volume. The NRMSE is given by equation 6.1. Here, the vector $\mathbf{f^{ref}} = \left(f_1^{ref}, ..., f_N^{ref}\right)^\mathbf{T} \in R^N$ is the software phantom volume used as a reference, the vector $\mathbf{f} = (f_1, ..., f_N)^\mathbf{T} \in R^N$ is the reconstructed volume from the simulated projection.

$$NRMSE = \sqrt{\sum_i \frac{\left(f_i - f_i^{ref}\right)^2}{\left(f_i^{ref} - \overline{f_i^{ref}}\right)^2}} \qquad (6.1)$$

The smaller NRMSE, the smaller the difference between the given reconstructed volume and the reference phantom volume.

6.1.3.2 Qualitative visual-based evaluation

Additionally, a qualitative visual-based inspection and comparison of the reconstructed slices has been carried out. Qualitative visual inspection is an important assessment, because it is difficult to capture the artifact specificity and to quantify the image quality using only one number (e.g. Bian 2010). The overall visual inspection and comparison of in-plane and two orthogonal axial slices has been performed with the attention on the four selected ROIs, shown in the Fig. 6.1.

6.2 Influence of the system acquisition parameters

The following acquisition parameters should be optimized in tomosynthesis: the angular range, the number of projections and and the angular step size. In each of the following subsections, first, a literature review on the impact of acquisition parameters to the performance of tomosynthesis and the clinical image quality will be given. Second, for each parameter, the state-of-the art conclusions will be compared with the simulation results in application to the imaging of hands.

6.2.1 The impact of the angular range θ

6.2.1.1 Literature review

One of the first studies about the impact of the angular range on the image quality in tomosynthesis was performed by Li et al. (Li 2004). The authors found, that an increase in the angular range leads to an increase in z-resolution. The overall relationship between the angular range and z-resolution was found to be non-linear with a linear behavior in a 20° to 40° range. The authors used a shallow-angled ramp phantom to measure the in-plane resolution defined by the MTF and to measure the slice thickness defined by the slice sensitivity profile (SSP) (Li 2006). The results of the study showed that an increase on the angular range has a small influence on the in-plane resolution but significantly reduces the slice thickness. Deller at al. (Deller 2007) performed a study using a resolution phantom and several anthropomorphic phantoms (chest, abdomen, wrist and pelvis) and reported similar conclusions: the angular range has almost no influence on the in-plane resolution and an increase in the angular range significantly increases the depth resolution. A number of texture features such as skewness, coarseness and contrast were used by Kontos et al. (Kontos 2008) to evaluate a reconstruction of a simulated anthropomorphic breast phantom. It was shown, that an increase in the angular range leads to more blurred out-of-focus artifacts and sharper in-focus structures. Sechopoulos et al. (Sechopoulos 2009) describes the in-plane lesion visibility and the axial resolution using the CNR and the ASF and included an empirical system response characteristics into the computer simulation study, which showed that an increase in the angular range always improves the axial resolution.

Using a three-dimensional cascaded linear system analysis with a realistic simulation of several parameters (MTF of the detector, a focal spot blur, the source-detector movement speed and the exposure) and a simulated thin tungsten wire and breast tissue with calcifications and a tumor, Zhou et al. (Zhou 2007) found that the MTF and NPS are mainly affected by the angular range. The second conclusion was, again,

6.2 Influence of the system acquisition parameters

that the in-depth resolution is improved when the angular range is increased. Another experiment showed that the MTF at low frequencies can be improved by increasing the angular range (Zhao 2008). Further studies using the linear system approach were performed by Hu et al. (Hu 2008a, Hu 2008b) who also belong to the abovementioned research group. The authors compared their model of the MTF and PSF with real measurements and concluded that an increase in the angular range narrows the PSF in z-direction and significantly reduces the intensity of the artifacts. Mertelmeier et al. (Mertelmeier 2008) also used the cascaded linear system model and reported that an increase in the angular range increases the axial resolution and improves the visibility of the low-frequency objects.

Tomosynthesis performance evaluation using an observer model-based measure of lesion detectability showed that the performance is improved when the angular range is increased (Chawla 2009). Investigation of tomosynthesis limitations in dependency of scan parameters and quantum noise based on mathematical observer models showed that the lesion detectability depends on the signal size (Reiser 2010). Increasing the angular range increases the detectability for all signals. In summary, the detectability is improved when the angular range is increased (Chawla 2009, Reiser 2010).

The recent works conclude that the imaging parameters should be adapted based on the visualized anatomy. For the visualization of large body parts as an abdomen, the angular range can be decreased, because the abdomen region does not contain fine structures. The depth resolution in this case is not critical. For the visualization of fine anatomy such as hands or feet, the angular range should be increased in order to improve depth resolution and reduce the risk of missing some subtle structures (Machida 2010).

One of the recent works in these area discusses the quantitative analysis of the imaging performance based on a simulation framework, a digital anthropomorphic breast phantom and a cascaded system analysis (CSA) for system modeling. The NPS has been used in this work to derive the figure of merits in terms of the accuracy and the precision of reconstruction for detection of lesion area, volume and location. Results show that decreasing angular range results in increased anatomical noise (out-of-focus artifacts). At the same time, usage of a larger angular range shows an increase in quantum and electronic noise. The precision performance was significantly dependent on the angular range, showing improved performance with larger angles, while accuracy was found to be independent of the angular range (Richard 2010b). One of the conclusions of their further work (Richard 2010a) is that imaging performance optimization should be based on the specific task. It might be challenging to optimize it for several different tasks at the same time.

6.2.1.2 Simulation results

The influence of the angular range has been studied for five fixed numbers of projections $N_{proj} = 6$, $N_{proj} = 12$, $N_{proj} = 25$, $N_{proj} = 50$ and $N_{proj} = 100$. The angular range has been varied from $\theta = \pm 5^o$ to $\theta = \pm 75^o$. Angles larger than $\pm 75^o$ become infeasible, because in the fixed detector geometry some parts of the object might be projected outside the detector area.

The NRMSE between the reconstructed volume and the reference volume in dependence of the angular range θ is presented in Fig. 6.3e. The curve with triangle markers describes $N_{proj} = 25$, which is used in clinical applications today. The overall curve behavior shows an improvement in the image quality when the angular range is increased. The improvement is higher for a large number of projections ($N_{proj} = 50 - 100$), than for a small number of projections ($N_{proj} = 6 - 12$). After a certain value of the angular range, the image quality decreases again. This can be seen in the curves corresponding to $N_{proj} = 6$ and $N_{proj} = 12$. Points of the NRMSE minimum are marked by short vertical bars. The angular range value corresponding to the minimum is equal $\pm 55^o$ for $N_{proj} = 6$ and $\pm 60^o$ for $N_{proj} = 12$. Increasing θ does not result in a dose increase if the number of projections is fixed. However, if the number of projections is insufficient, the image quality might decrease. Too large spacing between projections results in ripple artifacts because the blurring of out-of-focus structures does not work anymore.

The reconstruction results for $N_{proj} = 25$ and $\theta = \pm 12, \pm 25, \pm 50, \pm 75$ are shown in Fig. 6.3a-6.3d. The images illustrate the image quality improvement when increasing the angular range. The image obtained with $\theta = \pm 12^o$ (Fig. 6.3a) shows blurring artifacts in ROI1 and some lost bone boundaries in ROI2-ROI4. The image obtained with $\theta = \pm 75^o$ (Fig. 6.3a) shows great reduction of blur in the ROI1 and bone boundaries recovery in both, the in-plane ROI2 and axial ROI3 and ROI4. The general appearance of the image in Fig. 6.3a is very similar to the reference image (Fig. 6.1).

As a conclusion, the angular range θ plays an important role for the reconstructed image quality, which is increasing with increasing θ. At the same time, when the spacing between projections becomes too large, it results in the decrease of the image quality.

6.2 Influence of the system acquisition parameters

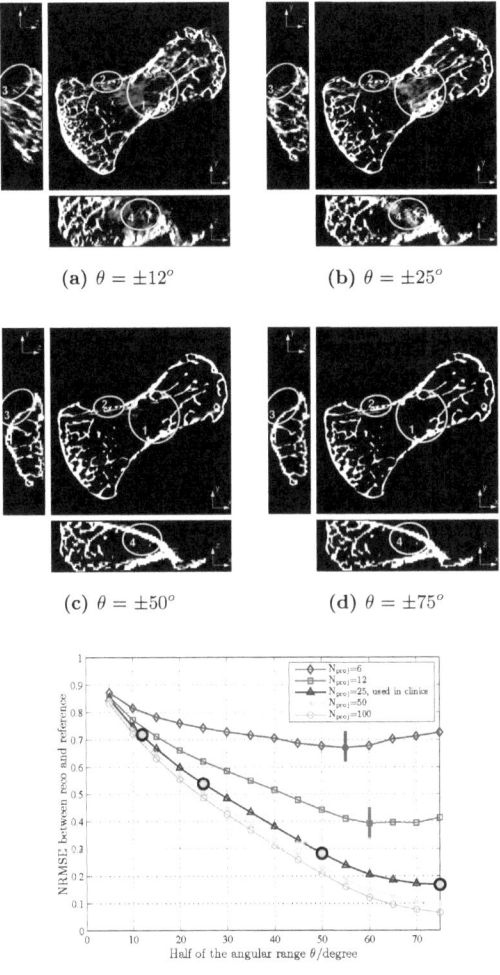

(a) $\theta = \pm 12^\circ$ (b) $\theta = \pm 25^\circ$

(c) $\theta = \pm 50^\circ$ (d) $\theta = \pm 75^\circ$

(e) NRMSE between the reconstructed volume and the reference volume

Figure 6.3: Simulation results showing the influence of the angular range θ. The reconstructed slices of the bone phantom for the fixed number of projections $N_{proj} = 25$ and varying θ are shown in (a-d). The corresponding points are marker by bold markers in the quantitative evaluation plot (e). The minimum of NRMSE for $N_{proj} = 6$ and $N_{proj} = 12$ is marked by a vertical bars. The values of the angular range, larger that those values results in the decrease in the image quality because of the ripple artifacts.

6.2.2 Influence of the number of projections N_{proj}

6.2.2.1 Literature review

An experimental and phantom study from Deller et al. (Deller 2007), showed that the number of projections has only minor influence on the resolution and image noise. At the same time, contrarily, Maidment et al. and Kontos et al. (Maidment 2006, Kontos 2008) concluded that an increasing in the number of projections results in image artifacts decrease, superior image quality and finer textures. However, maximizing the number of projections while keeping the angular range fixed does not change the appearance of the out-of-focus blurring (Kontos 2008). Ren et al. (Ren 2006) expect that in an ideal system an increase in the number of projections results in the reduction of out-of-focus artifacts. The authors performed a study with a tomosynthesis prototype system and used the CNR peak value and a profile shape as the image quality metrics. They concluded that adding more projections beyond a certain number, which depends on the angular range, is unnecessary. Going above this number does not further improve the image quality and axial resolution, but leads to an increase in the data size, processing time and dose. Sechopoulos et al. (Sechopoulos 2009) also agree that the image quality improvement effect in dependence of the number of projections tends to saturate.

Certain reconstruction algorithms, e.g. FBP and Matrix Inversion Tomosynthesis (MITS) perform better when the number of projections is increased (Chen 2005). An insufficient number of projections and large angular separation between views might lead to the ringing in the ASF, but it also depends on the reconstruction algorithm (Zhou 2007). Using less projections with the same angular range produces a similar ASF. However, in slices further away (5 mm) from in-focus slice of interest this results in an increase of ASF. A non-sufficient projection number results in the ripple artifacts (Hu 2008a).

Increasing the projection number while keeping the overall clinically relevant dose fixed means that a dose per exposure is decreased. It increases the noise and decrease the detectability and tomosynthesis performance (Sechopoulos 2009, Chawla 2009). Increasing the number of projections while keeping the dose per projection the same, i.e. increasing the total dose, decreases image noise (Machida 2010). Therefore, the dose and number of projections should be decreased for imaging thin objects such as hands or feet to avoid the unnecessary radiation dose (Machida 2010).

As a conclusion it can be said that the number of projections plays a less important role than the angular range. Improvement effects based on the larger projection number tend to saturate. The number of projections mainly contributes to the dose and the image noise but not to the appearance of the artifacts.

6.2 Influence of the system acquisition parameters

6.2.2.2 Simulation results

The influence of the number of projections has been studied for three fixed angular ranges with $\theta = \pm 12^o$, $\theta = \pm 25^o$ and $\theta = \pm 50^o$. The number of projections has been varied from $N_{proj} = 5$ to $N_{proj} = 100$. The NRMSE between the reconstructed volume and the reference volume in dependence on the number of projections is presented in Fig. 6.4e. Here, the curve with triangle markers corresponds to the angular range $\theta = \pm 25^o$, which is used in clinics and is equivalent to the Siemens Mammomat Inspiration geometry. The curves with diamond and circle markers describe the halved angular range value $\theta = \pm 12^o$ and the doubled angular range value $\theta = \pm 50^o$, correspondingly. The overall behavior of curves shows an improvement in image quality when the projection number is increased in the range of small number of projections. Above a certain threshold there is a plateau and a further increase in the number of projections does not result in a visible improvement. The saturation points are marked with short vertical bars. For example, the saturation value for $\theta = \pm 25^o$ is $N_{proj} = 25$.

Selected reconstruction results for $\theta = \pm 25^o$ and $N_{proj} = 3$, $N_{proj} = 12$, $N_{proj} = 25$ and $N_{proj} = 50$ are shown in Fig. 6.4a - Fig. 6.4d. The corresponding points are marked by bold markers in the evaluation plot Fig. 6.4e. The reconstructed images visually illustrate the behavior of NRMSE curves. First, the image obtained with $N_{proj} = 3$, shown in Fig. 6.4a, is blurry and contains only a few details. When the number of projections is increased to $N_{proj} = 12$ (Fig. 6.4b) and then to $N_{proj} = 25$ (Fig. 6.4c), more details become visible. However, no further improvement can be noticed when the number of projections is increased from $N_{proj} = 25$ to $N_{proj} = 50$ (Fig. 6.4d). The range of values $N_{proj} = 25$ to $N_{proj} = 50$ corresponds to the plateau region in the plot. In summary, when increasing the number of projections, the out-of-focus artifacts in the ROI1 are slightly reduced. The bone boundary in the in-plane ROI2 is better recovered, however the bone boundaries in the axial slices in ROI3 and ROI4 are still blurred and hardly recognizable. Additionally, in the clinical situation, the increase in number of measured projections means an increase in the total dose if the dose per projection is kept fixed. If the total dose is kept fixed it will result in more noise in each projection and, potentially, noisier reconstruction.

As a conclusion, an increase in the number of projections is feasible only up to a certain value, which depends on the angular range. This saturation value is between $N_{proj} = 15$ and $N_{proj} = 30$ for the wide range of angular range values from $\theta = \pm 12^o$ to $\theta = \pm 50^o$.

148   Chapter 6. Dual-axis tilt acquisition geometry

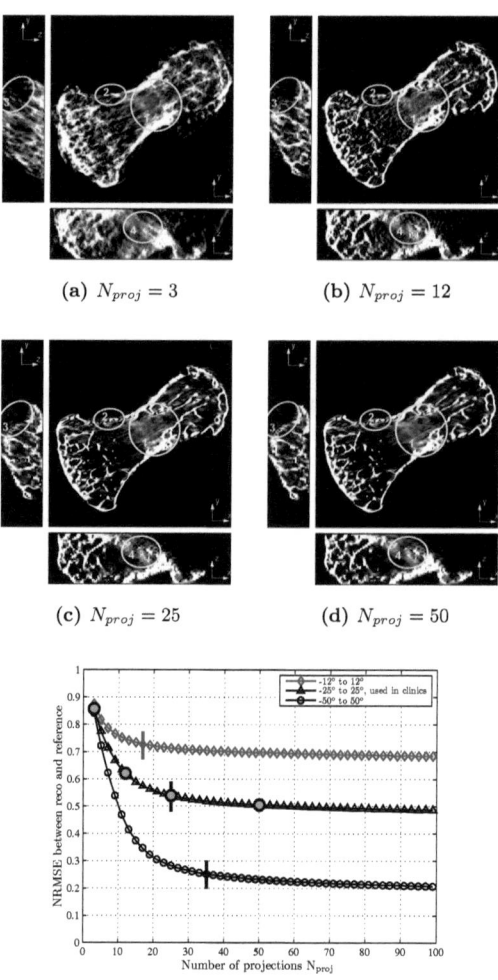

(a) $N_{proj} = 3$ (b) $N_{proj} = 12$

(c) $N_{proj} = 25$ (d) $N_{proj} = 50$

(e) NRMSE between the reconstructed volume
and the reference volume

Figure 6.4: Simulation results showing the influence of the number of projections N_{proj}. The reconstructed slices of the bone for the fixed angular range $\theta = \pm 25^o$ and varying the number of projections are shown in (a)-(d). The corresponding points are marked by bold markers in the quantitative evaluation plot (e). The points of NRMSE saturation are marked by short vertical bars. An increase of the number of projections larger than the saturation value does not increase the image quality further.

6.2.3 Influence of the angular step size $\Delta\theta$

6.2.3.1 Literature review

The angular step size is the relation between the angular range and the number of projections. The angular step size plays a more important role for artifact formation than the angular range and the number of projections separately. It is important to not only use a reasonable angular range value (e.g. $>30^o$) but also to limit the angular step size to small values (e.g. $<2^o$) (Hu 2008a).

Decreasing the angular step size with the fixed angular range helps to reduce ripple artifacts (Deller 2007), to avoid streak artifacts (Mertelmeier 2008) and to improve detectability (Reiser 2010). At the same time, the visibility of ripples also depends on the imaged anatomy and particularly on the thickness of the tissue and on the distance to the ripple-causing structures. There has been developed a formula for approximating a distance D from structure to the ripple artifact (Deller 2007)

$$D = N_{proj}/\left(2f \tan\left(\theta/2\right)\right). \quad (6.2)$$

Here, f is a cut-off frequency of the generalized filtered backprojection (GFBP). Therefore, the angular step size should be decreased when imaging a thick body part, such as chest with high-contrast objects (bones, ribs) to reduce the ripple artifacts. When a thick body part is examined, the angular step size can be increased. An increase in the value of angular range in this case will not lead to the formation of prominent ripple artifacts. (Machida 2010).

6.2.3.2 Simulation results

The influence of the angular step size has been studied for different combinations of the angular range from $\theta = \pm 5^o$ to $\theta = \pm 50^o$ and the number of projections from $N_{proj} = 5$ to $N_{proj} = 50$. It results in various angular step sizes from $\Delta\theta = 5^o$ to $\Delta\theta = 15^o$. The reconstructed results are shown in Fig. 6.5a - Fig. 6.5b. The dependence of the angular step size on the chosen pair (θ, N_{proj}) is shown in Fig. 6.5d. The NRMSE between the reconstructed volume and the reference volume for each chosen pair (θ, N_{proj}) is shown in Fig. 6.5d. The results are represented as contour lines. When following a contour line with fixed $\Delta\theta$, the image quality is increasing with increasing the angular range and the number of projections. The overall behavior of the curves summarizes the conclusions made in the previous two subsections. Increasing the number of projections with a fixed angular range, i.e. using a smaller angular step size, helps to improve the image quality for larger angular range values and has a saturation for the smaller angles. The angular

150 Chapter 6. Dual-axis tilt acquisition geometry

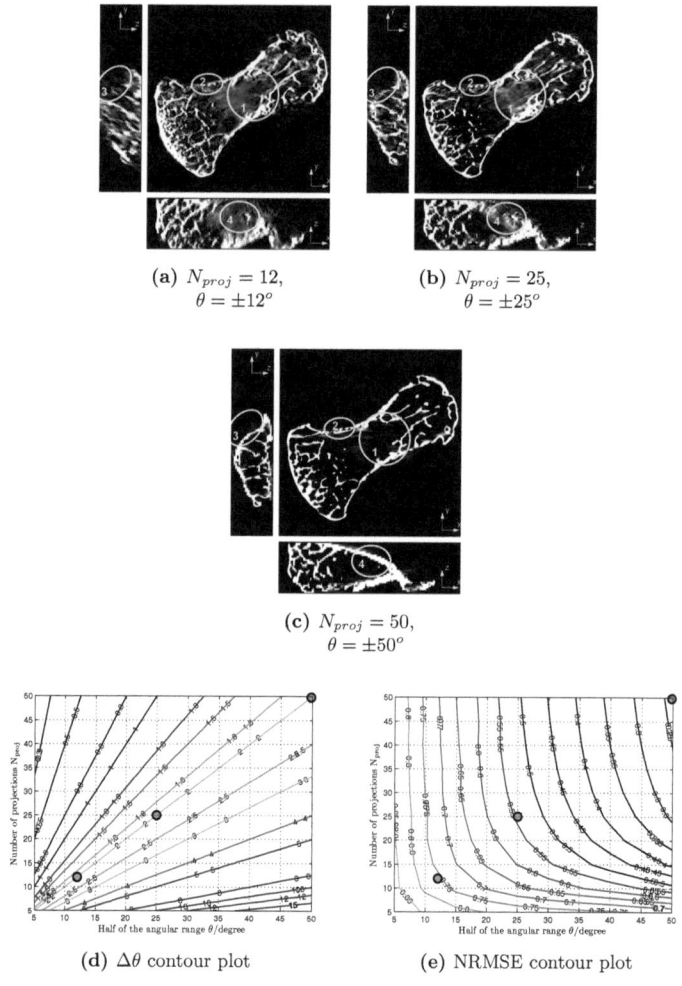

(a) $N_{proj} = 12$, $\theta = \pm 12°$

(b) $N_{proj} = 25$, $\theta = \pm 25°$

(c) $N_{proj} = 50$, $\theta = \pm 50°$

(d) $\Delta\theta$ contour plot

(e) NRMSE contour plot

Figure 6.5: Simulation results showing the influence of the angular step $\Delta\theta$. The reconstructed slices of the bone for the fixed angular step $\theta = 2°$ and varying θ and N_{proj} are shown in (a)-(c). The corresponding points are marked by bold markers in the quantitative evaluation plots (d) and (e).

range plays a more important role than the number of projections and an increase in the angular range with the fixed number of projections increases the image quality.

6.3 A novel geometry: hybrid dual-axis tilt acquisition

6.3.1 How to acquire more information of the object?

As it has been discussed in the previous section, the tomosynthesis performance can be improved by adjusting the geometry parameters. The main conclusion is that the more data about object is acquired, the better the obtained image quality. However, the image quality improvement is limited by newly introduced artifacts, e.g. by the ripple artifacts when the projection density is insufficient. This gives a motivation to search an alternative acquisition scheme, which can acquire more information of the object without introducing additional radiation dose or artifacts and, at the same time, provides images with less artifacts. Apparently, beside increasing the number of projections and the angular range, the acquisition geometry itself can be modified in order to acquire more data of the object.

There are three main geometry components, which can be modified: the detector orientation, the X-ray tube trajectory and the object orientation. An acquisition with a tilted detector does not result in additional data as it is shown in Fig. 6.6. Here, two projections, which differ from each other only by the detector orientation, can

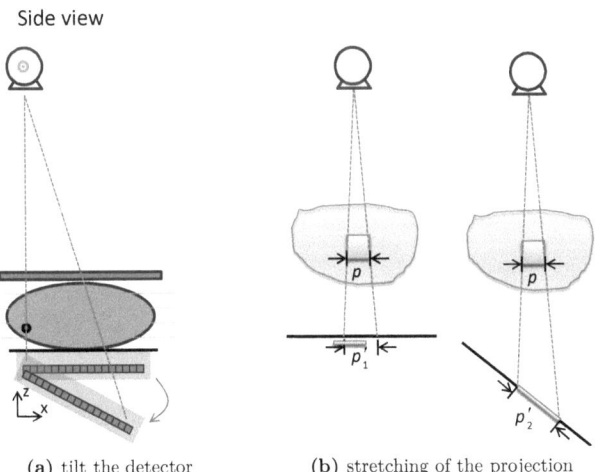

(a) tilt the detector   (b) stretching of the projection

Figure 6.6: Acquisition of the data by tilting the detector does not lead to an additional information of the object. Acquired projections can be transformed to each other by scaling and stretching operations.

be transformed into each other by simple scaling and stretching transformations. In theory, some improvements in the in-plane resolution can be gained (DiBianca 2000a, DiBianca 2000b, Dahi 2008) because small features will be projected onto a larger detector area when the detector is tilted, see Fig. 6.7b. Here, the feature of a size p is projected into p'_1 and p'_2 in case of no tilt and tilt, respectively. As it can be seen, p'_1 is smaller than p'_2, $p'_1 \leq p'_2$. This kind of image resolution improvement will not be discussed in this work, since it does not reduce artifacts.

Moving the X-ray tube over a two-dimensional trajectory with all other geometry parameters being fixed allows for capturing the data from the sides (Zhang 2010) and thus it improves the image quality. At the same time, it will result in additional patient dose when the X-ray tube moves in the direction away (Fig. 6.7a) or toward (Fig. 6.7b) to the patient. Additionally, a re-design of existing tomosynthesis devices would be needed for moving the tube in the orthogonal direction. Moving or rotating the object

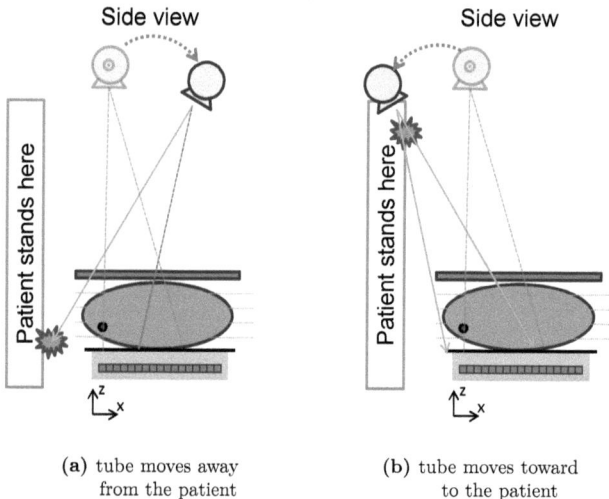

(a) tube moves away from the patient

(b) tube moves toward to the patient

Figure 6.7: Acquisition of the data by tilting the X-ray tube might cause additional radiation dose to the patient.

as it is done, e. g. in the micro-CT, is the third possibility to acquire more data. In the tomosynthesis case the limitations occur because the human body and its parts cannot be freely moved at any position and orientation.

6.3 A novel geometry: hybrid dual-axis tilt acquisition

6.3.2 Theoretical background

A novel acquisition geometry is proposed, which has been inspired by a solid angle tomosynthesis (Zhang 2010) and tomosynthesis with an array of X-ray tubes (Maltz 2009, Qian 2012). In the solid angle tomosynthesis, the acquisition is done by positioning the X-ray source in two perpendicular directions. Alternatively, specially designed compact multiple source X-ray tubes based on carbon nanotube field-emission cold cathode technology (CNT) can be combined into a one-dimensional or two-dimensional array. A similar dual axis acquisition approach is also known from the scanning transmission electron microscopy (STEM) (Iancu 2005, Arslan 2006). In this case an object is tilted in two perpendicular directions using a tiltable stage.

The hybrid approach addressed in this chapter is a composition of methods mentioned above. The acquisition in one direction is still done by moving the tube along the standard arc trajectory. The acquisition in the additional perpendicular direction is done by tilting the object. The standard existing device configuration is extended using an additional object-support tiltable platform, mounted on the detector. The platform tilts the object in y-direction orthogonal to the X-ray tube rotation (x-axis). Instead of taking one image, several images are measured at each X-ray tube position, differing only in the tilt angle of the platform, see Fig. 6.8.

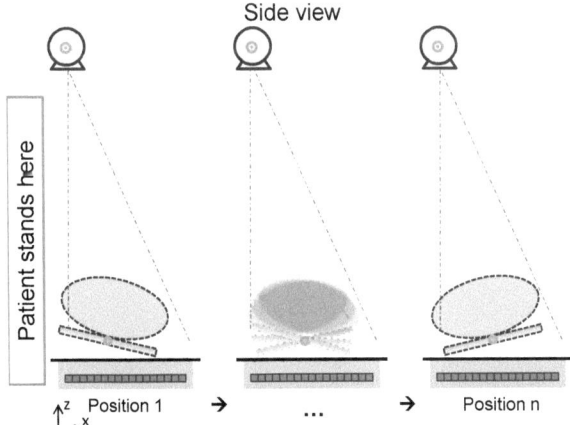

Figure 6.8: A hybrid dual-axis tiltable acquisition geometry. Acquisition in original direction is done by moving the tube, while acquisition in the perpendicular direction is done by tilting the object.

Tilting the object is geometrically equivalent to moving the X-ray tube outside the standard arc trajectory. This way, for each position of the X-ray tube several projections outside the standard arc can be acquired without additional movement of the tube in the perpendicular direction. The total number of projections and, therefore, the applied dose stays the same, because the number of projections in the original direction is reduced.

The proposed geometry allows for capturing more singularities of the Radon transform (Quinto 1993), filling more data in the Fourier Space and better approximating the Tuy-Smith Conditions (Tuy 1983, Smith 1985). These three theoretical aspects of tomographic imaging have been discussed in detail in Chapter 2. In the next subsection, they will be discussed with respect to the novel geometry.

6.3.3 Singularities of Radon transform and limited data

A singularity of the object is a boundary between two tissues. Singularities of the object are closely related to the singularities of the Radon transform. A singularity of the object produces a singularity in the Radon transform only if there is a line integral measured in the direction tangential to this singularity. Such singularities are called *visible* and can be stably reconstructed. In the three-dimensional case, the X-ray tube at each position casts a *visible* ring on a sphere object. The orientation of the visible ring depends on the xy position and height z of the sphere. An arc acquisition trajectory AB results in a set of visible rings, see Fig. 6.9a. All rings crosses in two opposite points. All other points on the sphere, marked by dashed lines remain invisible and cannot be reconstructed stably. If the X-ray tube is moved to the point C, which is located in the perpendicular plane, a visible ring C_1C_2 with an essentially new orientation will be produced, see Fig. 6.9b. To summarize, the dual-axis acquisition geometry allows for capturing singularities which are located in perpendicular planes (i.e. in the plane of new virtual X-ray tube movement) and would have stayed invisible in case of arc trajectory.

6.3.4 Incomplete Fourier space

According to the well-known Fourier slice theorem, the limited angle tomosynthesis acquisition along an arc θ_x covers only the limited wedge θ_x in the Fourier domain. The dual-axis tilt acquisition allows for filling an additional wedge in the xz-plane. A schematic representation of the measurement process and the Fourier space are shown in Fig. 6.10.

6.3 A novel geometry: hybrid dual-axis tilt acquisition 155

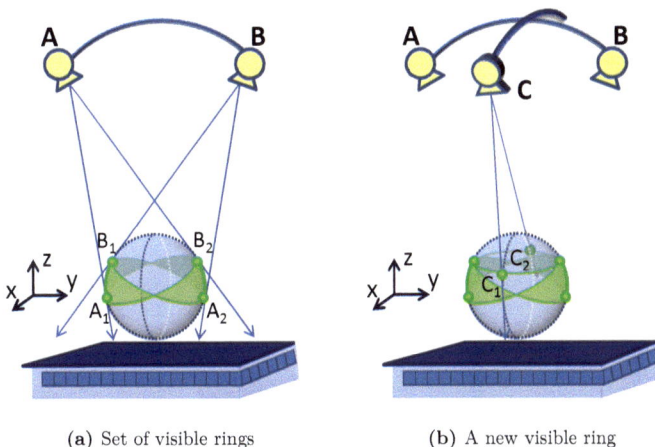

(a) Set of visible rings for the acquisition along AB

(b) A new visible ring casted in the position C

Figure 6.9: (a) An acquisition along the one-dimensional arc results in a set of *visible* rings of singularities. (b) If the X-ray tube is moved to the point C, which is located in the perpendicular plane, a visible ring C_1C_2 with an essentially new orientation is produced.

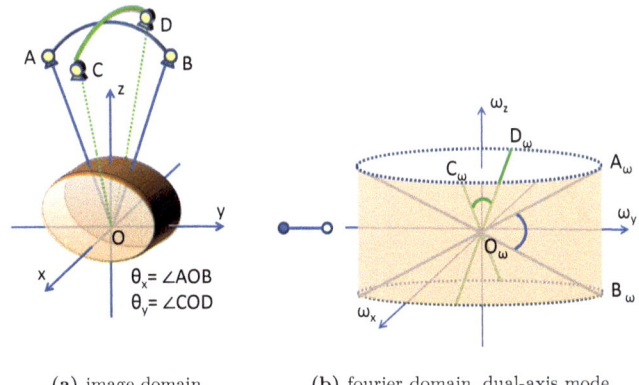

(a) image domain

(b) fourier domain, dual-axis mode

Figure 6.10: Illustration of the incomplete Fourier domain. The wedge $\theta_x = \angle AOB = (\angle AOB)_\omega$ corresponds to the standard acquisition mode. The wedges $\theta_x = \angle AOB = (\angle AOB)_\omega$ and $\theta_y = \angle COD = (\angle COD)_\omega$ correspond to the dual-axis tilt acquisition.

The arc AB and the wedge $\theta_x = \angle AOB = (\angle AOB)_\omega$ correspond to the standard acquisition mode. The additional arc CD and the wedge $\theta_y = \angle COD = (\angle COD)_\omega$ correspond to the dual-axis tilt acquisition. With the new geometry not only a wedge is filled in the Fourier space, but a cylinder with a missing cone along the z-direction. It shows that more data become available, which can be used for reconstruction.

6.3.5 Tuy-Smith sufficiency condition

The Tuy-Smith sufficiency condition (Tuy 1983, Smith 1985) states that if on every plane that intersects the object there exists at least one cone-beam source point, then one can reconstruct the object exactly. The standard tomosynthesis trajectory does not fulfill this condition. With additional data acquired outside the standard arc trajectory, more planes through the object intersects the X-ray tube trajectory (see Fig. 6.11b). Therefore, the degree of violation of the Tuy-Smith sufficiency condition is decreased.

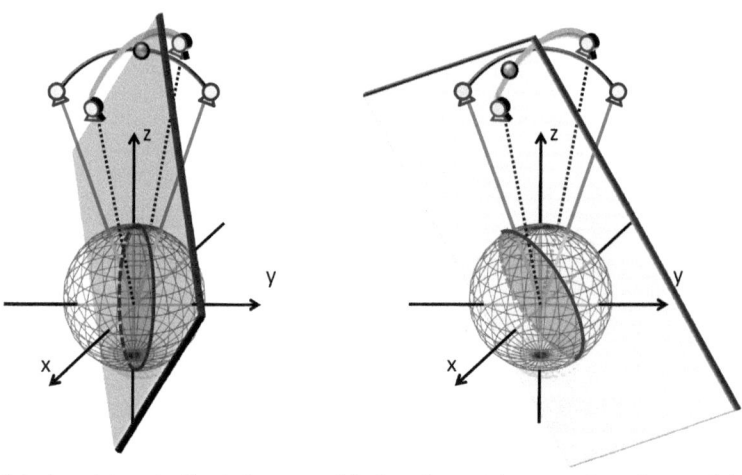

(a) plane intersects the trajectory of the standard 1D-arc geometry; condition fulfilled

(b) plane does not intersects the trajectory of the 1D-arc geometry, but intersects the second arc of the tiltable geometry; condition fulfilled

Figure 6.11: Illustration of two planes which fulfill the Tuy Smith sufficiency conditions in standard mode and in tiltable mode.

6.3.6 Angle in x-direction and re-distribution of projections.

The influence of the angle in x-direction has been studied using an angular range in the original direction $\theta_y = \pm 25^o$ and the fixed number of projections. The angle in x-direction has been varies from $\theta_x = \pm 2^o$ to $\theta_x = \pm 25^o$. A fixed number of projections can be re-distributed in two perpendicular directions in several fashions. If the number of projections $N_{proj} = 24$, they can be re-distributed as following: $N_{proj,y}$ =6 and $N_{proj,x} = 4$, $N_{proj,y} = 8$ and $N_{proj,x} = 3$, $N_{proj,y} = 12$ and $N_{proj,x} = 2$. Additionally, if the root of the total number of projection is a natural number, they can be redistributed "symmetrically" $N_{proj,x} = 5$, $N_{proj,y} = 5$. We are aware of the fact that for the "symmetrical" pattern N_{proj} is 25. However, we would like to test all possible patterns and based on the experiments with standard geometry presented in the previous section, we assume that the results obtained using $N_{proj} = 24$ and $N_{proj} = 25$ are fairly comparable with each other. For the comparison, the standard acquisition geometry has been simulated using $\theta = \pm 25^o$, and $N_{proj} = 25$ based on the geometry of the Siemens Mammomat Inspiration. The simulation results are presented in Fig. 6.12e. THe horizontal line with o-markers corresponds to the standard mode with no additional angle in x-direction. When the value of θ_x is small ($\pm 2^o$) and the $N_{proj,x} = 2$, the dual axis geometry performs slightly better than the standard geometry. When the number of projections in x-direction is increased, the image quality of the dual-axis mode becomes even worse than the one standard geometry. It happens because an increase of $N_{proj,x}$ means decrease of $N_{proj,y}$, which, in turn leads to the decrease in image quality. This effect can be clearly seen in the reconstructed images in Fig. 6.12a and Fig. 6.12b. In both images the bone boundaries in ROI2-ROI4 are lost and out-of-focus blurring is strongly present. All curves in the plot show the main tendency of improving the image quality when larger θ_x is used for all projection re-distribution schemes. When θ_x is relatively large ($\pm 25^o$) and it is value is equal θ_y, the "symmetrical" re-distribution scheme performs the best, compare Fig. 6.12c and Fig. 6.13d. In both images the bone boundaries in ROI2-ROI4 are recovered, however in Fig. 6.12c one can see the noticeable residual blurring in ROI1. At the same time, multiple tilting steps of the object over the wide angular range might be practically difficult.

158 Chapter 6. Dual-axis tilt acquisition geometry

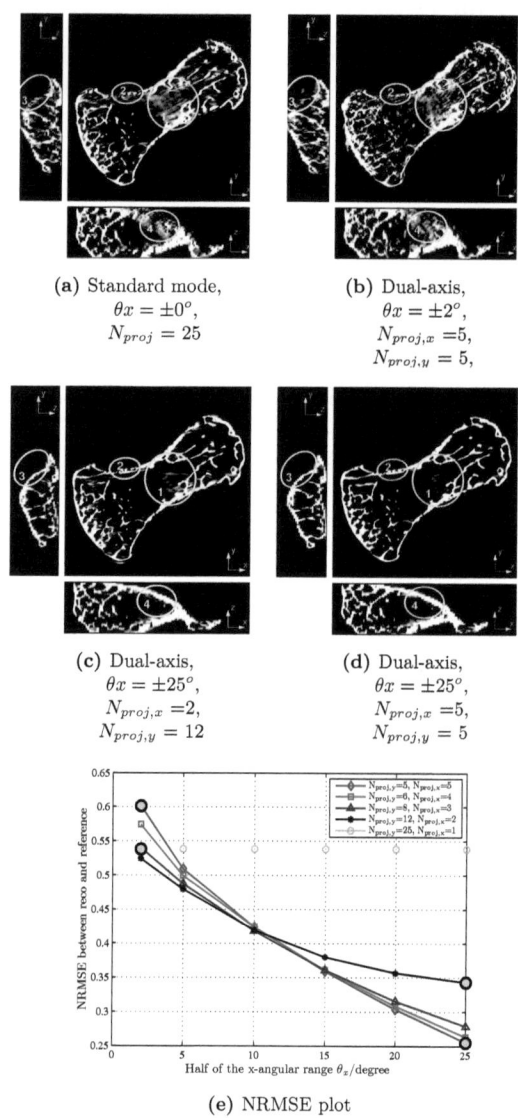

(a) Standard mode,
$\theta x = \pm 0°$,
$N_{proj} = 25$

(b) Dual-axis,
$\theta x = \pm 2°$,
$N_{proj,x} = 5$,
$N_{proj,y} = 5$,

(c) Dual-axis,
$\theta x = \pm 25°$,
$N_{proj,x} = 2$,
$N_{proj,y} = 12$

(d) Dual-axis,
$\theta x = \pm 25°$,
$N_{proj,x} = 5$,
$N_{proj,y} = 5$

(e) NRMSE plot

Figure 6.12: Simulation results for varying angular range in x-direction and different re-distribution of projections for dual axis geometry. The reconstructed slices of the finger bone (a)-(d) are highlighted with bold markers in the quantitative evaluation plot (e). The ○-markers correspond to the standard mode with no angle in x-direction.

6.3.7 Influence of number of projections

To evaluate the influence of number of projections, the best performance parameters from the previous experiment (previous subsection) have been taken. The dual-axis tilt mode has been simulated using $\theta_x = \theta_y = \pm 25^o$ and the symmetrical projection re-distribution. Four different numbers of projections have been tested. The following values have been selected $N_{proj} = 16, 25, 36, 49$, because their roots are natural numbers. In the symmetrical projection re-distribution scheme the number of projection per each axis in the tilt geometry is a root of the total number of projections in the standard geometry. Therefore, the corresponding dual-axis tilt mode has been simulated using $\theta_x = \theta_y = \pm 25^o$ and $N_{proj,x} = N_{proj,y} = 4, 5, 6, 7$ which is equivalent to a total projection number N_{proj} of 16, 25, 36, 49 in standard mode

$$N_{proj} = N_{proj,x} \times N_{proj,y}. \tag{6.3}$$

The simulation results are presented in Fig. 6.13e. The line with diamond markers corresponds to data using standard acquisition mode and the line with triangle markers correspond to data gained in dual-axis tilt acquisition mode. The NRMSE is much smaller when using the dual-axis tilt acquisition mode in all tested cases compared to the standard acquisition mode. The image quality is improved when more projections are used for both acquisition modes.

The reconstruction results with $N_{proj} = 16$ and 25 in standard mode are presented in Fig. 6.13a and 6.13c. This is equivalent to $N_{proj,y} = N_{proj,y} = 4$ and 5 in the dual-axis mode. The corresponding images are presented in Fig. 6.13b and 6.13d. The images illustrate notable image improvement when using dual-axis mode. The images obtained with standard acquisition mode (Fig. 6.13b, Fig. 6.13c) are affected by blurring due to out-of focus artifacts in ROI1. Bone boundaries in in-plane ROI2 and axial ROI3 and ROI4 are lost. The corresponding images obtained with the tiltable platform (Fig. 6.13b, Fig. 6.13d) show less out-of-focus artifacts. The bone boundaries are recovered in ROI2–ROI4.

As a conclusion, using the same number of projections (i.e. the fixed total dose), the dual-axis geometry provides images with better quality compared to the standard acquisition mode. The out-of-focus artifacts are reduced and the in-plane and axial boundaries of the bone are restored.

Chapter 6. Dual-axis tilt acquisition geometry

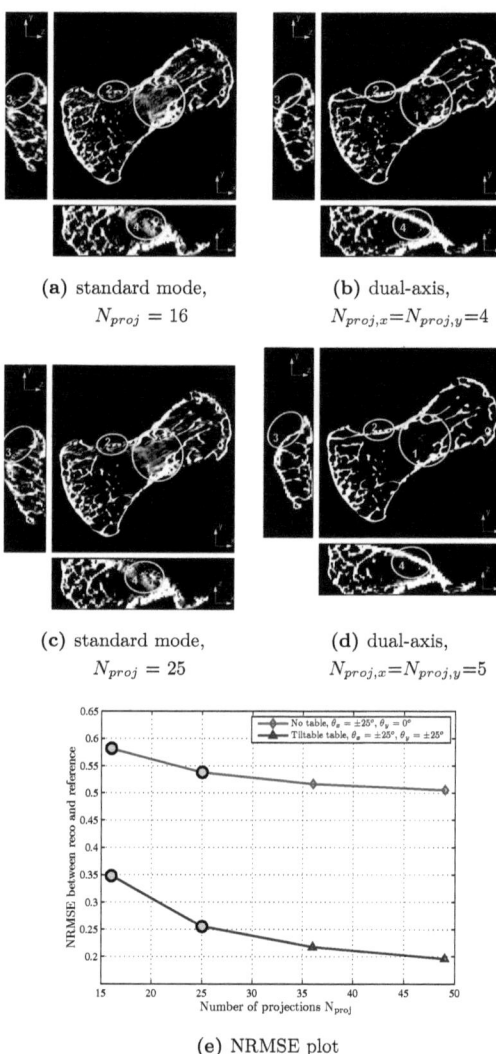

(a) standard mode, $N_{proj} = 16$

(b) dual-axis, $N_{proj,x} = N_{proj,y} = 4$

(c) standard mode, $N_{proj} = 25$

(d) dual-axis, $N_{proj,x} = N_{proj,y} = 5$

(e) NRMSE plot

Figure 6.13: Simulation results for the standard acquisition and the acquisition with the tiltable platform. The reconstructed slices of the bone are shown in (a)-(d). The NRMSE plot is shown in (e). The standard mode parameters are: $\theta_x = \pm 25°$, $\theta_y = 0$. The dual-axis mode parameters are: $\theta_x = \pm 25°$, $\theta_y = \pm 25°$.

6.4 Influence of the object orientation

It is known that the sweep direction should be chosen by taking into account the thickness of the measured body part and the particular purposes of the study (Cordes 2011, Cordes 2012a, Cordes 2012b, Cordes 2013). The image quality depends on the in-plane object orientation with respect to the tube arc direction. Depending on the object orientation, some features can appear distorted (Machida 2010).

The influence of the object orientation with respect to the X-ray tube rotation axis has been studied for the standard mode and the dual-axis acquisition mode. The standard acquisition mode parameters are: $\theta = \pm 25°$, $N_{proj} = 25$. The dual-axis acquisition mode parameters are: $\theta_x = \theta_y = \pm 25°$, $N_{proj,x} = N_{proj,y} = 5$. The NRMSE curves measured for varying object orientations from $0°$ to $360°$ and two acquisition modes are presented in Fig. 6.14.

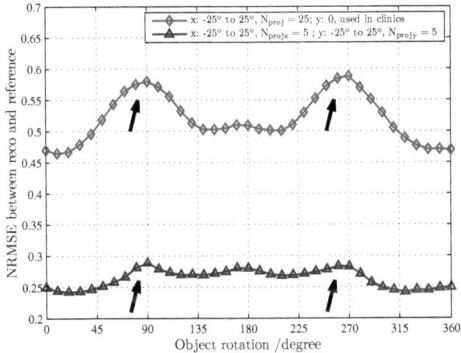

Figure 6.14: Varying the object orientation with respect to the X-ray tube rotation for the standard and dual-axis tilt acquisition modes.

The curve corresponding to the standard mode is marked by diamond markers and the curve corresponding to the dual axis mode is marked by triangle markers. In the standard acquisition mode the image quality decreases when the orientation angle is close to $90°$ or $270°$. The loss of image quality manifests itself as peaks in Fig. 6.14, indicated by arrows. The dual-axis acquisition mode does not show those peaks, and, therefore, reduces the dependency of image quality on the object orientation. In addition, the image quality for the dual-axis acquisition mode is better than in the standard mode for all object orientations.

Chapter 7

Conclusions and suggestions for further work

This doctoral thesis addresses the three-dimensional imaging modality called digital tomosynthesis. It is an X-ray based limited angle tomographic technique for the visualization of inner structures of an object with the advantage of high in-plane resolution and low radiation dose. However, the data acquired along the limited angular range are incomplete, which leads to artifacts in reconstructed images. The main goal of this work is to improve tomosynthesis image quality by reducing the limited angle artifacts. Based on the literature review, theoretical aspects and experimental studies, it has been shown that this can be achieved either by using an improved reconstruction technique or by using a more appropriate acquisition geometries. Two main contributions of this work are a weighted reconstruction algorithm and a dual-axis acquisition geometry.

In chapter 2, a review on tomosynthesis has been presented. It starts with an introduction of basic principles and a historical development of the technology and the reconstruction algorithms and includes a comparison of a state of the art tomosynthesis device with CT and micro-CT devices in terms of technical parameters and reconstructed images. The historical overview showed that tomosynthesis is a technology with a rich and interesting history and which is strongly connected to CT. The comparison of modern clinical CT, micro-CT and tomosynthesis devices showed that tomosynthesis is an attractive alternative to the three-dimensional X-ray based imaging techniques. It has been shown that tomosynthesis provides images with high in-plane resolution and that the limited in-depth resolution of tomosynthesis images does not prevent structures of being reconstructed at their correct locations. As the second part of chapter 2, the theoretical aspects of the image reconstruction from the limited data have been discussed in terms of the Radon and Fourier transforms and the Tuy-Smith sufficiency condition. If these assumptions and conditions are not fulfilled, this results in images artifacts. Understanding of the underlying theory is important because it allows for a deep insight into the artifacts formation, which, in turn, is important to understand for the development of artifact reduction strategies. For future work, a more detailed study on the artifact formation in tomosynthesis should be done. It might include the artifact formation in dependency on certain shapes and tissue types of the artifact-causing features, an influence of reconstruction algorithms, and device parameters (e.g. X-ray tube, detector). The visualization of the reconstructed dataset (slice-by-slice, volume rendering) is also an important aspect, which influences the subjective impression caused by artifacts to observers and can be a subject of further studies. Additionally, non-medical application of tomosynthesis with the two-dimensional circular trajectory can be also considered for further studies.

In the third chapter, the forward- and backprojection algorithms, which are the key part of any reconstruction algorithm, are discussed. The discussion includes the object discretization using series expansion for a practical implementation, the choice of the basis functions (pixel, blobs) with their properties and an evaluation model for line integral calculation. Based on the literature review and practical experience, the state of the art distance driven-projection algorithm has been chosen for the reconstruction framework. An efficient practical implementation strategy for the distance-driven algorithm has been proposed. It has been described in detail for pixels and blobs basis functions for the two-dimensional fan-beam CT. The angular cases of the distance-driven algorithm have been formulated as a single function without redundant "if-else" construction and a sweep line principle has been used. The implementation has been further extended for the three-dimensional tomosynthesis geometry. It has been shown how the implementation of this algorithm in three-dimensions can benefit from the fixed detector tomosynthesis geometry. Future work should include the implementation of the algorithms within a C++ reconstruction framework and further acceleration using a GPU. In general, the GPU-based implementation of any type of forward and backprojections becomes more and more favorable (see e.g. the proceedings of the 3rd workshop on High Performance Image Reconstruction (HPIR) (HPIR 2011)). Further improvements might include the modeling of physical effects such as scatter, beam hardening, the polyenergetic spectrum, the focal spot size, the detector efficiency, etc. Additional questions are if tomosynthesis can benefit from a projector, which models the pixel footprints more accurate or from the usage of overlapping basis functions (blobs, b-splines). However, it is important to remember that there is always the trade-off

Chapter 7. Conclusions and further work 165

between a realistic acquisition model and the speed of the projectors, which in turn defines the speed of iterative reconstruction algorithms.

In chapter 4, the reconstruction problem from projections has been formulated as an optimization problem. Two types of iterative reconstruction algorithms have been mentioned: the algebraic and the statistical reconstruction. The main focus of this chapter is the simultaneous algebraic reconstruction technique (SART) for tomosynthesis. The discussion includes practical implementation details and the influence of the projection access order parameter. A novel data-based projection access order has been proposed, which minimizes total correlation between projections. This order has been compared with several projection access order schemes known from CT theory and adapted for tomosynthesis. The simulation study showed that in the limited angle case the projection order influences the convergence rate and might lead the iteration scheme to different solutions. The minimum correlation approach seem to have great potential, however, a number of further improvements and experiments are required. This includes the development of a path search algorithm, which is suitable for this particular problem and further simulation and real-data studies. Additionally, other iterative and non-iterative algorithms, which are outside the scope of this work may also offer many opportunities for further investigations.

In chapter 5, the backprojected space (BP) representation has been proposed and its properties have been discussed. A novel dissimilarity concept for tomosynthesis in BP-space has been proposed. It was shown how dissimilarity can be used to construct a weighting scheme, which can be included either in the simple backprojection or in the algebraic reconstruction algorithm resulting in the non-linear ωBP and ωSART methods, correspondingly. The weighting coefficients individually control the contribution of each ray to each voxel in the reconstructed image and thus reduce the tomosynthesis artifacts produced by high-absorption features. Preliminary reconstruction results of real hand datasets confirm that the proposed weighting scheme reduces the out-of-focus artifacts. Additionally, an application of BP-space representation for metal artifact reduction in CT has been discussed, which includes an extension of the ωSART algorithm for the CT geometry and a usage of the BP-space to follow a sinogram flow for the interpolation. More information on the relation between the dissimilarity and the weighting and an optimization of the weighting curve parameters would be helpful to achieve better results. Further research might explore the usage of the BP-space for the metal artifact problem in CT in more details and considerably more work will need to be done using simulated and real data.

In contrast to chapters 3 - 5 where only reconstruction methods have been considered for the tomosynthesis artifact reduction, in chapter 6, the acquisition geometry and parameters have been addressed. The impact of the acquisition parameters on the tomosynthesis image quality has been studied using a simulation study and a realistic bone phantom. This study has shown a good agreement with conclusions, which can be found in literature on breast tomosynthesis. A novel dual-axis tilt acquisition geometry based on a tiltable platform has been proposed. It has been proposed to acquire the data outside the standard one-dimensional arc trajectory, which is typically used in clinical practice. The proposed geometry allows for capturing more singularities of the Radon transform, fill more data in the Fourier space and provide more planes, which satisfy the Tuy-Smith condition. The simulation results showed that the dual-axis tilt acquisition principle offer to reach better in-plane and axial image quality using the same number of projections (i.e. the same dose) without major modification of the existing tomosynthesis devices. The studies described in this chapter were based on a simulation software. Results confirm the feasibility of the proposed acquisition geometry. However, further work should be done in order to explore all potentials and limitations.

The parameters of the novel acquisition trajectory should be optimized. The main question is how to efficiently re-distribute the acquired projections over the new trajectory. Another aspect, which should be taken into account, is the influence of the reconstruction algorithm. The task is to find a suitable reconstruction algorithm for the limited angle data acquired over non-standard trajectory. Further efforts should be made in an investigation of the potential problems of the motion artifacts introduced by movement of the object. As an important next step, a hardware prototype should be designed and implemented. Further validation studies based on real measured data, which includes the qualitative and quantitative evaluation should be done. A validation based on task-specific image quality measures has not been discussed in this chapter. Therefore, it is important to carry out such studies. The patient dose issue has only been slightly mentioned in this chapter. The proposed geometry offer not only image quality improvement without patient dose increase but also a potential of the patient dose reduction while keeping image quality constant.

To sum up, the proposed geometry concept would benefit from investigation studies in the following areas:

- optimizing the geometry parameters;
- finding a suitable reconstruction algorithm for non-standard trajectories;
- investigating potential motion artifacts;
- studying dose issues;
- building a prototype;
- demonstrating the scheme on the real data.

Chapter 8

Appendix: MATLAB®

8.1 MATLAB® File Exchange

The following tools from the MATLAB® file exchange (webMATLAB 2013) have been used to improve the visual representation of graphics created in MATLAB®:

- `varycolor` for maximum color variation for lines on plots;
- `legend best fit` to position automatically the existing legend inside the axes to avoid covering the plotted data;
- `legendflex` for legend with more flexible positioning and labelling capabilities;
- `ellipse` to add ellipses to the current plot;
- `save2pdf` to save a figure as a PDF as it appears on the screen with the correct page size;
- `arrow` to draw a line with an arrowhead.

Additionally, the `mcode` package has been used to include m-code fragments with the highlighted syntax in the LaTeX documents.

8.2 PubMed trend search

The source-code represents a MATLAB® function for PubMed trend search. An alternative tool for PubMed trends search be found online here (Corlan 2004).

```
1  function trend=get_pubmed_trends(years, search_term)
2  %function get_pubmed_trends returns occurance of published papers per year for ...
       given search_term
3  %
4  %the code is adapted from
5  %http://www.vollkornpapier.de/allgemeines/...
6  %...trends-fur-publikationen-zu-einem-thema-in-pubmed.html
7  %
8  %Example: trend=get_pubmed_trends(1972:2012, 'tomosynthesis')
9  %26/09/2012
10
11  trend=zeros(length(years),1); % memory preallocation for output
12  for j=1:length(years)
13      query=['http://eutils.ncbi.nlm.nih.gov/entrez/eutils/ ...
14          esearch.fcgi?db=pubmed&term=' urlencode(search_term) ...
15          '[tiab]%20AND%20' num2str(years(j)) '[dp]&rettype=count'];
16      docNode = xmlread(query);
17      trend(j)=str2double(docNode.getElementsByTagName('Count').item(0). ...
18          getFirstChild.getNodeValue);
19  end
```

Chapter 9

References

[Ambrose 1973] J. Ambrose. *Computerised transverse axial scanning (tomography): Part 2. Clinical application.* Br. J. Radiol., vol. 46, pages 1023–1047, 1973. 2, 21

[Andersen 1984] A. H. Andersen & A. C Kak. *Simultaneous algebraic reconstruction technique (SART): a superior implementation of the ART algorithm.* Ultrason. Imaging, vol. 6, no. 1, pages 81–94, 1984. 78

[Andersen 1989] A. H. Andersen. *Algebraic reconstruction in CT from limited views.* IEEE Trans. Med. Imag., vol. 8, no. 1, pages 50–55, 1989. 78

[Arslan 2006] I. Arslan, J. R. Tong & P. A. Midgley. *Reducing the missing wedge: high-resolution dual axis tomography of inorganic materials.* Ultramicroscopy, vol. 106, no. 11-12, pages 994–1000, 2006. 153

[Badea 1998] C. Badea, Z. Kolitsi & N. Pallikarakis. *A wavelet-based method for removal of out-of-plane structures in digital tomosynthesis.* Comput. Med. Imaging Graph., vol. 22, no. 4, pages 309–315, 1998. 24

[Baker 2011] J. A. Baker & J. Y. Lo. *Breast tomosynthesis: state-of-the-art and review of the literature.* Acad. Radiol, vol. 18, no. 10, pages 1298–1310, 2011. 12

[Beckmann 2006] E. C. Beckmann. *CT scanning the early days.* British Journal Of Radiology, vol. 79, no. 937, pages 5–8, 2006. 29th Annual Presdents Conference, Arlington, VA, 30.Jan-01.Feb, 2005. 21

[Bevilacqua 2007] R. Bevilacqua, E. Bozzo & O. Menchi. *Comparison of four natural pixel bases for SPECT imaging.* Journal Of Computational And

Applied Mathematics, vol. 198, no. 2SI, pages 361–377, 2007. Workshop on Applied Computational Inverse Problems, CNR, Florence, Italy, 22-25 Mar, 2004. 48

[Bian 2010] J. Bian, J. H. Siewerdsen, X. Han, E. Y. Sidky, J. L. Prince, C. A. Pelizzari & X Pan. *Evaluation of sparse-view reconstruction from flat-panel-detector cone-beam CT*. Phys. Med. Biol., vol. 55, no. 22, pages 6575–6599, 2010. 141

[Bippus 2011] R. D. Bippus, T. Koehler, F. Bergner, B. Brendel, E. Hansis & R. Proksa. *Projector and backprojector for iterative CT reconstruction with blobs using CUDA*. In 11th International Meeting on Fully Three-Dimensional Image Reconstruction in Radiology and Nuclear Medicine, Potsdam, Germany, pages 68–71, 2011. 56, 57

[Birkfellner 005] W. Birkfellner, R. Seemann, M. Figl, J. Hummel, C. Ede, P. Homolka, X. H. Yang, P. Niederer & H. Bergmann. *Wobbled splatting - a fast perspective volume rendering method for simulation of x-ray images from CT*. Phys. Med. Biol., vol. 50, no. 9, pages N73–N84, 2005. 47

[Bocage 1822] A. E. M. Bocage. *Procede et dispositif de radiographie sur plaque en mouvement*, 1822. French Paten FR 236464. 19

[Bracewell 1956] R. N. Bracewell. *Strip Integration in Radio Astronomy*. Australian Journal of Physics, vol. 9, pages 198–216, 1956. 37

[Buonocore 1981] M. H Buonocore, W. R. Brody & A. Macovski. *A natural pixel decomposition for two-dimensional image-reconstruction*. IEEE Trans. Biomed. Eng., vol. 28, no. 2, pages 69–78, 1981. 48

[Buzug 2008] T. M. Buzug. Computed tomography: From photon statistics to modern cone-beam CT. Springer-Verlag, Berlin/Heidelberg, 2008. 38

[Candes 1999] E. J. Candes & D. L. Donoho. *Ridgelets: a key to higher-dimensional intermittency?* Philos. Trans. R. Soc. Lond. Ser. A-Math. Phys. Eng. Sci., vol. 357, no. 1760, pages 2495–2509, 1999. 101

[Caramelo 2005] F. J. Caramelo, N. C. Ferreira, L. Fazendeiro & C. Souto. *Image reconstruction by sinogram decomposition into sinusoidal curves*. In 8th International Meeting on Fully Three-Dimensional Image Reconstruction in Radiology and Nuclear Medicine, pages 55–58, 2005. 100

REFERENCES 171

[Carvalho 2003] B. M. Carvalho & G. T. Herman. *Helical CT reconstruction from wide cone-beam angle data using ART*. In Computer Graphics and Image Processing, 2003. SIBGRAPI 2003. XVI Brazilian Symposium on, pages 363–370, 2003. 49

[Chakraborty 1984] D. P. Chakraborty, M. V. Yester, G. T. GBarnes & A. V. Lakshminarayanan. *Self-masking subtraction tomosynthesis*. Radiology, vol. 150, no. 1, pages 225–229, 1984. 23

[Chawla 2009] A. S. Chawla, J. Y. Lo, J. A. Baker & E. Samei. *Optimized image acquisition for breast tomosynthesis in projection and reconstruction space*. Med. Phys., vol. 36, pages 4859–4869, 2009. 143, 146

[Chen 2005] Y. Chen, J. Y. Lo & J. T. Dobbins. *Impulse response analysis for several digital tomosynthesis mammography reconstruction algorithms*. In Proc. SPIE, volume 5745, pages 541–549. San Diego, CA, USA, 2005. 146

[Chlewicki 2004] W. Chlewicki, F. Hermansen & S. B. Hansen. *Noise reduction and convergence of Bayesian algorithms with blobs based on the Huber function and median root prior*. Phys. Med. Biol., vol. 49, no. 20, pages 4717–4730, 2004. 49

[Christiaens 1999] M. Christiaens, B. De Sutter, K. De Bosschere, J. Van Campenhout & I. Lemahieu. *A fast, cache-aware algorithm for the calculation of radiological paths exploiting subword parallelism*. Journal Of Systems Architecture, vol. 45, no. 10, pages 781–790, 1999. 48

[Claus 2006] B. Claus, J. Eberhard, A. Schmitz, P. Carson, M. Goodsitt & H. Chan. *Generalized filtered back-projection reconstruction in breast tomosynthesis*. In Susan Astley, Michael Brady, Chris Rose & Reyer Zwiggelaar, editeurs, Digital Mammography, volume 4046 of *Lecture Notes in Computer Science*, pages 167–174. Springer Berlin / Heidelberg, 2006. 25

[Cordes 2011] A. Cordes. *Evaluation of tomosynthesis reconstruction based on micro-CT*, November 2011. Bachelor thesis, University of Luebeck, Institute of Medical Engineering. 17, 161

[Cordes 2012a] A. Cordes, Y. M. Levakhina & T. M. Buzug. *A method for validation and evaluation of digital tomosynthesis reconstruction*. In Biomed Tech, page 513, 2012. 17, 161

REFERENCES

[Cordes 2012b] A. Cordes, Y. M. Levakhina & T. M. Buzug. *Mikro-CT basierte Validierung digitaler Tomosynthese Rekonstruktion.* In Bildverarbeitung für die Medizin, pages 304–309, 2012. 17, 161

[Cordes 2013] A. Cordes, Y. M. Levakhina & T. M. Buzug. *A method for validation and evaluation of digital tomosynthesis reconstruction.* Phys. Med. Biol., vol. submitted, pages –, 2013. 17, 161

[Corlan 2004] A. D. Corlan. *Medline trend: automated yearly statistics of PubMed results for any query.* http://dan.corlan.net/medline-trend.html, 2004. [Online; accessed 20-February-2013, Archived by WebCite at http://www.webcitation.org/65RkD48SV]. 168

[Cormack 1963] A. M. Cormack. *Representation of a function by its line integrals, with some radiological applications.* J. Appl. Phys., vol. 34, no. 9, pages 2722–2727, 1963. 33, 34

[Cormack 1964] A. M. Cormack. *Representation of a function by its line integrals, with some radiological applications, II.* J. Appl. Phys., vol. 35, no. 9, pages 2908–2913, 1964. 33, 34

[Cormack 1979] A. M. Cormack. *Early two-dimensional reconstruction and recent topics stemming from it,* 1979. Nobel Lecture, 8 December. 34

[Dahi 2008] B. Dahi, G. S. Keyes, D. A. Rendon & F. A. DiBianca. *Analysis of axial spatial resolution in a variable resolution x-ray cone beam CT (VRX-CBCT) system.* In Proc. SPIE, volume 6913, pages 69134Y–1, 2008. 152

[De Man 2002] B. De Man & S. Basu. *Distance-driven projection and backprojection.* In Nuclear Science Symposium Conference Record, IEEE, volume 3, pages 1477 – 1480, 2002. 48

[De Man 2003] B. De Man & S. Basu. *Distance-driven projection and backprojection.* In 7th International Meeting on Fully Three-Dimensional Image Reconstruction in Radiology and Nuclear Medicine, pages 199–202, 2003. 48

[De Man 2004] B. De Man & S. Basu. *Distance-driven projection and backprojection in three dimensions.* Phys. Med. Biol., vol. 49, no. 11, pages 2463–2475, 2004. 48, 62, 140

[De Man 2005] B. De Man, S. Basu, J.-B. Thibault, J. A. Jiang H.and Fessier, C. Bouman & K. Sauer. *A study of four minimization approaches for iterative reconstruction in X-ray CT.* In Nuclear Science Symposium Conference Record, IEEE, volume 5, pages 2708–2710, 2005. 77

REFERENCES

[Deller 2007] T. Deller, K. N. Jabri, J. M. Sabol, X. Ni, G. Avinash, R. Saunders & R. Uppaluri. *Effect of acquisition parameters on image quality in digital tomosynthesis.* In Proc. of SPIE, volume 6510, pages 65101L1–11, 2007. 17, 142, 146, 149

[DiBianca 2000a] F. A. DiBianca, V. Gupta & H. D. Zeman. *A variable resolution X-ray detector for computed tomography: I. Theoretical basis and experimental verification.* Med. Phys., vol. 27, no. 8, pages 1865–1874, 2000. 152

[DiBianca 2000b] F. A. DiBianca, P. Zou, L. M. Jordan, J. S. Laughter, H. D. Zeman & J. Sebes. *A variable resolution X-ray detector for computed tomography: II. Imaging theory and performance.* Med. Phys., vol. 27, no. 8, pages 1875–1880, 2000. 152

[Dobbins 2003] J. T. Dobbins & D. J. Godfrey. *Digital x-ray tomosynthesis: current state of the art and clinical potential.* Phys. Med. Biol., vol. 48, no. 19, pages R65–R106, 2003. 16, 21, 25

[Dobbins 2008] J. T. Dobbins, H. P. McAdams, J. Song, C. M. Li, D. J. Godfrey, D. M. DeLong, S. Paik & S. Martinez-Jimenez. *Digital tomosynthesis of the chest for lung nodule detection: Interim sensitivity results from an ongoing NIH-sponsored trial.* Med. Phys., vol. 35, no. 6, pages 2554–2557, 2008. 12

[Dobbins 2009] J. T. Dobbins. *Tomosynthesis imaging: At a translational crossroads.* Med. Phys., vol. 36, no. 6, pages 1956–1967, 2009. 21

[Duryea 2003] J. Duryea, J. T. Dobbins & J. A. Lynch. *Digital tomosynthesis of hand joints for arthritis assessment.* Med. Phys., vol. 30, no. 3, pages 325–333, 2003. 12

[Edholm 1969] P. R. Edholm & L. Quiding. *Reduction of linear blurring in tomography.* Radiology, vol. 92, no. 5, pages 1115–1118, 1969. 23

[Edholm 1980] P. Edholm, G. Granlund, H. Knutsson & C. Petersson. *Ectomography - a new radiographic method for reproducing a selected slice of varying thickness.* Acta Radiol. Diagn., vol. 21, no. 4, pages 433–442, 1980. 24

[Entezari 2012] A. Entezari, M. Nilchian & M. Unser. *A Box spline calculus for the discretization of computed tomography reconstruction problems.* IEEE Trans. Med. Imag., vol. 31, no. 8, pages 1532–1541, 2012. 45

[Erdogan 1999] H. Erdogan & J. A. Fessler. *Ordered subsets algorithms for transmission tomography.* Phys. Med. Biol., vol. 44, no. 11, pages 2835–2851, 1999. 81

[Faridani 2003] A. Faridani. *Introduction to the mathematics of computed tomography*. Inside Out: Inverse Problems and Application, vol. 47, pages 1–46, 2003. 35

[Feldkamp 1984] L. A. Feldkamp, L. C. Davis & J. W. Kress. *Practical cone-beam algorithm*. J. Opt. Soc. Am., vol. 1, no. 6, pages 612–619, 1984. 138

[Fessler 1997a] J. Fessler. *Equivalence of pixel-driven and rotation-based backprojectors for tomographic image reconstruction*. http://web.eecs.umich.edu/~fessler/papers/lists/files/reject/97,tmi,eop.pdf, 1997. [Online; accessed 18-January-2013]. 47

[Fessler 1997b] J. A. Fessler, E. P. Ficaro, N. H. Clinthorne & K. Lange. *Grouped-coordinate ascent algorithms for penalized-likelihood transmission image reconstruction*. IEEE Trans. Med. Imag., vol. 16, no. 2, pages 166–175, 1997. 81

[Fessler 1999] JA Fessler & SD Booth. *Conjugate-gradient preconditioning methods for shift-variant PET image reconstruction*. IEEE Trans. Med. Imag., vol. 8, no. 5, pages 688–699, 1999. 77

[Fessler 2000] J. Fessler. *Statistical image reconstruction for transmission tomography*. In Milan Sonka J. Michael Fitzpatrick, editeur, Handbook of medical imaging, volume 2, pages 1 – 70. SPIE Press Book, Bellingham WA, 2000. 77, 81

[Galigekere 2003] R. R. Galigekere, K. Wiesent & D. W. Holdsworth. *Cone-beam reprojection using projection-matrices*. IEEE Trans. Med. Imag., vol. 22, no. 10, pages 1202–1214, 2003. 48

[Gao 2012] H. Gao. *Fast parallel algorithms for the X-ray transform and its adjoint*. Med. Phys., vol. 39, no. 11, pages 7110–7120, 2012. 48

[Garduno 2004] E. Garduno & G. T. Herman. *Optimization of basis functions for both reconstruction and visualization*. Discrete Applied Mathematics, vol. 139, no. 1-3, pages 95–111, 2004. International Workshop on Combinatorial Image Analysis (IWCIA 2001), Philadelphia, PA, 23-24 Aug, 2001. 49, 55

[Garrison 1969] J. B. Garrison, D. G. Grant, W. H. Guier & R. J. Johns. *Three dimensional roentgenography*. Am. J. Roentgenol. Radium. Ther. Nucl. Med., vol. 105, no. 4, pages 903–908, 1969. 20, 23

[Gilbert 1972] P. Gilbert. *Iterative methods for three-dimensional reconstruction of an object from projections*. J. Theor. Biol., vol. 36, no. 1, pages 105–117, 1972. 78

REFERENCES

[Godfrey 2006] D. J. Godfrey, H. P. McAdams & J. T. Dobbins. *Optimization of the matrix inversion tomosynthesis (MITS) impulse response and modulation transfer function characteristics for chest imaging.* Med. Phys., vol. 33, no. 3, pages 655–667, 2006. 24

[Gordon 1970] R. Gordon, R. Bender & G. T. Herman. *Algebraic reconstruction techniques (ART) for 3-dimensional electron microscopy and X-ray photography.* J. Theor. Biol, vol. 29, no. 3, pages 478–481, 1970. 78

[Gould 2009] R. G. Gould. *Early days of CT: innovations (both good and bad).* http://www.aapm.org/meetings/amos2/pdf/42-12236-29343-41.pdf, 2009. [Online; accessed 31-October-2012]. 21

[Grant 1972] D. G. Grant. *Tomosynthesis: a three-dimensional radiographic imaging technique.* IEEE Trans. Biomed. Eng., vol. BM19, pages 20–28, 1972. 16, 21

[Grossmann 1935a] G Grossmann. *Tomographie 1: Röntgenographische Darstellung von Körperschnitten.* Forschr. Röntgenstr., vol. 51, pages 61–80, 1935. 19

[Grossmann 1935b] G Grossmann. *Tomographie 2: Theoretisches über Tomographie.* Forschr. Röntgenstr., vol. 51, pages 191–208, 1935. 19

[Guan 1994] H. Q. Guan & R. Gordon. *A projection access order for speedy convergence of art (algebraic reconstruction technique) - a multilevel scheme for computed-tomography.* Phys. Med. Biol., vol. 39, no. 11, pages 2005–2022, 1994. 87, 93

[Guan 1996] H. Q. Guan & R. Gordon. *Computed tomography using algebraic reconstruction techniques (ARTs) with different projection access schemes: a comparison study under practical situations.* Phys. Med. Biol., vol. 41, no. 9, pages 1727–1743, 1996. 87, 88

[Guan 1998] H. Q. Guan, R. Gordon & Y. P. Zhu. *Combining various projection access schemes with the algebraic reconstruction technique for low-contrast detection in computed tomography.* Phys. Med. Biol., vol. 43, no. 8, pages 2413–2421, 1998. 87, 88

[Guedouar 2010] R. Guedouar & B. Zarrad. *A comparative study between matched and mis-matched projection/back projection pairs used with ASIRT reconstruction method.* Nuclear Instruments & Methods In Physics Research Section A-accelerators Spectrometers Detectors And Associated Equipment, vol. 619, no. 1-3, pages 225–229, 2010.

11th International Symposium on Radiation Physics, Melbourne, Australia, 20-25 Sep, 2009. 48

[Gullberg 1985] G. T. Gullberg, R. H. Huesman, J. A. Malko, N. J. Pelc & T. F. Budinger. *An attenuated projector backprojector for iterative SPECT reconstruction.* Phys. Med. Biol., vol. 30, no. 8, pages 799–816, 1985. 48

[Haaker 1985a] P. Haaker, E. Klotz, R. Koppe, R. Linde & D. G. Mathey. *First clinical results with digital flashing tomosynthesis in coronary angiography.* Eur. Heart. J., vol. 6, no. 11, pages 913–920, 1985. 24, 109

[Haaker 1985b] P. Haaker, E. Klotz, R. Koppe, R. Linde & H. Moller. *A new digital tomosynthesis method with less artifacts for angiography.* Med. Phys., vol. 12, no. 4, pages 431–436, 1985. 24, 109

[Hanson 1985] K. M. Hanson & G. W. Wecksung. *Local basis-function approach to computed tomography.* Applied Optics, vol. 24, no. 23, pages 4028–4039, 1985. 44

[Happonen 2002] A. P. Happonen & S. Alenius. *Sinogram filtering using a stackgram domain.* In Proceedings of the Second IASTED International Conference: Visualization, Imaging and Image Processing, pages 339–343, 2002. 100

[Happonen 2003] A. P. Happonen & U. Ruotsalainen. *Alignment of scans in dynamic pet study using stackgram domain.* In Nuclear Science Symposium Conference Record (NSS/MIC), IEEE, pages 2868–2872, 2003. 100

[Happonen 2004] A. P. Happonen & U. Ruotsalainen. *Three-dimensional alignment of scans in a dynamic PET study using sinusoidal trajectory signals of a sinogram.* IEEE Trans. Nucl. Sci., vol. 51, no. 5, Part 2, pages 2620–2627, 2004. 100

[Happonen 2005a] A. Happonen. *Decomposition of Radon projections into stackgrams for filtering, extrapolation, and alignment of sinogram data.* PhD thesis, Tampere University of Technology, November 2005. Publication 557. 100

[Happonen 2005b] A. P. Happonen & S. Alenius. *A comparison of sinogram and stackgram domain filtering methods employing L-filters for noise reduction of tomographic data.* In Proceedings of the 2005 Finnish Signal Processing Symposium (FINSIGt05), pages 1–4, 2005. 100

REFERENCES

[Happonen 2005c] A. P. Happonen & U. Ruotsalainen. *A comparative study of angular extrapolation in sinogram and stackgram domains for limited angle tomography*. In Lecture Notes in Computer Science, volume 3540, pages 1047–1056, 2005. 100

[Happonen 2007a] A. P. Happonen & M. O. Koskinen. *Experimental investigation of angular stackgram filtering for noise reduction of SPECT projection data: study with linear and nonlinear filters*. Int. J. Biomed. Imaging., vol. 2007, pages 1–12, 2007. Article ID 38516. 100

[Happonen 2007b] A. P. Happonen & M. O. Koskinen. *Preliminary results on noise reduction using stackgrams for (low-dose) X-ray CT sinograms*. In Proceedings of the 2007 European Signal Processing Conference (EUSIPCO-2007), pages 1093–1097, 2007. 100

[Harauz 1983] G. Harauz & F. P. Ottensmeyer. *Interpolation in computing forward projections in direct three-dimensional reconstruction*. Phys. Med. Biol., vol. 28, no. 12, pages 1419–1427, 1983. 47

[Herman 1993] G. T. Herman & L. B. Meyer. *Algebraic reconstruction techniques can be made computationally efficient*. IEEE Trans. Med. Imag., vol. 12, no. 3, pages 600–609, 1993. 87, 89

[Hounsfield 1973] G. N. Hounsfield. *Computerised transverse axial scanning (tomography): Part 1. Description of system*. Br. J. Radiol., vol. 46, pages 1016–1022, 1973. 2, 21

[HPIR 2011] HPIR. In Proceedings of the 3rd Workshop on High Performance Image Reconstruction, 2011. [Online; accessed 20-January-2013]. 164

[Hu 2008a] Y. H. Hu, B. Zhao & W. Zhao. *Image artifacts in digital breast tomosynthesis: Investigation of the effects of system geometry and reconstruction parameters using a linear system approach*. Med. Phys, vol. 35, pages 5242–5251, 2008. 39, 143, 146, 149

[Hu 2008b] Y. H. Hu, W. Zhao, T. Mertelmeier & J. Ludwig. *Image artifact in digital breast tomosynthesis and its dependence on system and reconstruction parameters*. Digital Mammography, pages 628–634, 2008. 143

[Huang 2004] Z. Huang, Z. Zheng Li & K. Kang. *The application of digital tomosynthesis to the CT nondestructive testing of long large objects*. In Proc. SPIE, volume 5535, 2004. http://dx.doi.org/10.1117/12.558695. 12

REFERENCES

[Huesmanand 1977] R. H. Huesmanand, G. T. Gullberg, W. L. Greenberg & T. F Budinger. *RECLBL library users manual donner algorithms for reconstruction tomography.* ftp://cfi.lbl.gov/pub/reclbl/man/reclbl_man.pdf, 1977. [Online; accessed 18-January-2013]. 46

[Iancu 2005] C. V. Iancu, E. R. Wright, J. Benjamin, W. F. Tivol, D. P. Dias, G. E. Murphy, R. C. Morrison, J. B. Heymann & G. J. Jensen. *A "flip-flop" rotation stage for routine dual-axis electroncryotomography.* J. Struct. Biol., vol. 151, no. 3, pages 288–297, 2005. 153

[Isola 2008] A. A. Isola, A. Ziegler, T. Koehler, W. J. Niessen & M. Grass. *Motion-compensated iterative cone-beam CT image reconstruction with adapted blobs as basis functions.* Phys. Med. Biol., vol. 53, pages 6777–6797, 2008. 49

[Jacobs 1998] F. Jacobs, E. Sundermann, B. De Sutter, M. Christiaens & I. Lemahieu. *A fast algorithm to calculate the exact radiological path through a pixel or voxel space.* Journal of Computing and Information Technology, vol. 6, pages 89–94, 1998. 48

[Jacobs 1999a] F. Jacobs & I. Lemahieu. *Iterative image reconstruction from projections based on generalised Kaiser-Bessel window functions.* In Proceedings of the 1999 International Meeting on Fully Three-Dimensional Image Reconstruction in Radiology and Nuclear Medicine, pages 43–46, 1999. 49, 52

[Jacobs 1999b] F. Jacobs, S. Matej, R. M. Lewitt & I. Lemahieu. *A comparative study of 2D reconstruction algorithms using pixels and optimized blobs applied to Fourier rebinned 3D data.* In Proceedings of the 6th World Congress on Industrial Process Tomography (WCIPT), pages 14–17, 1999. 49

[Jiang 2003a] M. Jiang & G. Wang. *Convergence of the simultaneous algebraic reconstruction technique (SART).* IEEE Trans. Img. Proc., vol. 12, no. 8, pages 957–961, 2003. 79

[Jiang 2003b] M. Jiang & G. Wang. *Convergence studies on iterative algorithms for image reconstruction.* IEEE Trans. Med. Imag., vol. 22, no. 5, pages 569–579, 2003. 79

[Joseph 1982] P. M. Joseph. *An improved algorithm for reprojecting rays through pixel images.* IEEE Trans. Med. Imag., vol. MI-1, no. 3, pages 192–196, 1982. 48

REFERENCES

[Kaczmarz 1937] S. Kaczmarz. *Angenäherte Auflösung von Systemen linearer Gleichungen*. Bull. Internat. Acad. Polon. Sci. Lettres A,, vol. 35, pages 355–357, 1937. 78

[Kak 1984] A. C. Kak. *Image reconstruction from projections*. In MP Ekstorm, editeur, Digital image processing techniques, pages 111–170. Academic Press, Orlando FL, 1984. 77

[Kak 1988] A. Kak & M. Slaney. *Principles of computerized tomographic imaging*. IEEE Press, New York, 1988. 79

[Kieffer 1929] J. Kieffer. *X-ray device and method of technique*, 1929. US Patent No 1954321. 19

[Kiencke 2013a] S. Kiencke. *Data-based projection access order for SART*, March 2013. Master thesis, University of Luebeck, Institute of Medical Engineering. 95, 96

[Kiencke 2013b] S. Kiencke, Y. M. Levakhina & T. M. Buzug. *Greedy projection access order for SART (Simultaneous Algebraic Reconstruction Technique)*. In Bildverarbeitung für die Medizin, pages 93–98, 2013. 95

[Knutsson 1980] H. E. Knutsson, P. Edholm, G. H. Granlund & C. U. Petersson. *Ectomography - a new radiographic reconstruction method. 1. Theory and error estimates*. IEEE Trans. Biomed. Eng., vol. 27, no. 11, pages 640–648, 1980. 24

[Kohler 2004] T. Kohler. *A projection access scheme for iterative reconstruction based on the golden section*. In Nuclear Science Symposium Conference Record, IEEE, volume 6, pages 3961–3965, 2004. 87, 89

[Köhler 2011] T. Köhler, B. Brendel & E. Roessl. *Iterative reconstruction for differential phase contrast imaging using spherically symmetric basis functions*. Med. Phys., vol. 38, pages 4542–4545, 2011. 49

[Kolitsi 1993] Z. Kolitsi, G. Panayiotakis & N. Pallikarakis. *A method for selective removal of out-of-plane structures in digital tomosynthesis*. Med. Phys., vol. 20, no. 1, pages 47–50, 1993. 24

[Kong 2011] H. Kong & J. Pan. *A projection access scheme based on prime number increment for cone-beam iterative reconstruction*. In Advanced Electrical and Electronics Engineering, volume 87 of *Lecture Notes in Electrical Engineering*, pages 179–185. Springer Berlin Heidelberg, 2011. 87, 93

REFERENCES

[Kong 2012] H. Kong & J. Pan. *Evaluation of the projection ordering for cone-beam CT*. Journal of Computational Information Systems (JCIS), vol. 8, no. 10, pages 4031–4038, 2012. 87

[Kontos 2008] C. Kontos D.and Zhang, N. Ruiter, P. Bakic & A. Maidment. *Evaluating the effect of tomosynthesis acquisition parameters on image texture: a study based on an anthropomorphic breast tissue software model*. In Digital Mammography, volume 5116 of Lecture Notes in Computer Science, pages 491–498, 2008. 142, 146

[Kostopoulos 2006] A. E. Kostopoulos, A. P. Happonen & U. Ruotsalainen. *The 3-D alignment of objects in dynamic PET scans using filtered sinusoidal trajectories of sinogram*. Nuclear Instruments & Methods In Physics Research Section A-accelerators Spectrometers Detectors and Associated Equipment, vol. 569, no. 2SI, pages 434–439, 2006. 3rd International Conference on Imaging Technologies in Biomedical Sciences, Milos, Greece, 25-29 Sep, 2005. 100

[Krestyannikov 2004a] E. Krestyannikov, A. P. Happonen & Ruotsalainen U. *Noise models for sinusoidal trajectories composing sinogram data in positron emission tomography*. In Proceedings of the 6th Nordic Signal Processing Symposium (NORSIG 2004), pages 77–80, 2004. 100

[Krestyannikov 2004b] E. Krestyannikov & U. Ruotsalainen. *Quantitatively accurate data recovery from attenuation-corrected sinogram using filtering of sinusoidal trajectory signals*. In Nuclear Science Symposium Conference Record (NSS/MIC), IEEE, pages 3195–3199, 2004. 100

[Lange 1984] K. Lange & R. Carson. *EM reconstruction algorithms for emission and transmission tomography*. J. Comp. Assisted Tomo., vol. 8, no. 2, pages 306–316, 1984. 81

[Lauritsch 1998] G. Lauritsch & W. H. Haerer. *Theoretical framework for filtered back projection in tomosynthesis*. Proc. SPIE, vol. 3338, pages 1127–1137, 1998. 25

[Levakhina 2010] Y. M. Levakhina & T. M. Buzug. *Distance driven projection and backprojection for spherically symmetric basis functions*. In Nuclear Science Symposium Conference Record, IEEE, pages 2894–2897, 2010. 56

[Levakhina 2011a] Y. M. Levakhina, R. L. Duschka, J. Barkhausen & T. M. Buzug. *Digital tomosynthesis of hands using simultaneous algebraic reconstruction technique with distance-driven projector*. In 11th

REFERENCES 181

International Meeting on Fully Three-Dimensional Image Reconstruction in Radiology and Nuclear Medicine, pages 167–170, 2011. 25, 140

[Levakhina 2011b] Y. M. Levakhina, B. Kratz, R. L. Duschka, F. Vogt, J. Barkhausen & T. M. Buzug. *Reconstruction for musculoskeletal tomosynthesis: a comparative study using image quality assessment in image and projection domain.* In Nuclear Science Symposium Conference Record, IEEE, pages 2569–2571, 2011. 25

[Levakhina 2012a] Y. M. Levakhina, R. L. Duschka, F. M. Vogt, J. Barkhausen & T. M. Buzug. *An adaptive spatially-dependent weighting scheme for tomosynthesis reconstruction.* In Proc. SPIE, volume 8313, pages 831350–831356, 2012. 24, 100, 106, 111

[Levakhina 2012b] Y. M. Levakhina, R. L. Duschka, F. M. Vogt, J. Barkhausen & T. M. Buzug. *A novel acquisition scheme for higher axial resolution and improved image quality in digital tomosynthesis.* In Biomed Tech, volume 57, pages 111–114, 2012. 137

[Levakhina 2012c] Y. M. Levakhina, J. Mueller, R. L. Duschka, F. M. Vogt, J. Barkhausen & T. M. Buzug. *Algebraic tomosynthesis reconstruction with spatially adaptive updating term.* In Proceedings of The Second International Conference on Image Formation in X-Ray Computed Tomography, pages 46–49, 2012. 100, 106

[Levakhina 2013a] Y. M. Levakhina, R. L. Duschka, F. M. Vogt, J. Barkhausen & T. M. Buzug. *A hybrid dual-axis tilt acquisition geometry for digital musculoskeletal tomosynthesis.* Phys. Med. Biol., vol. submitted, pages 1–14, 2013. 137

[Levakhina 2013b] Y. M. Levakhina, J. Mueller, R. L. Duschka, F. M. Vogt, J. Barkhausen & T. M. Buzug. *Weighted simultaneous algebraic reconstruction technique for tomosynthesis imaging of objects with high-attenuation features.* Med. Phys., vol. 40, pages 1–12, 2013. 100, 120

[Lewitt 1990] R. M. Lewitt. *Multidimensional digital image representations using generalized Kaiser-Bessel window functions.* J. Opt. Soc. Am. A, vol. 7, no. 10, pages 1834–1846, 1990. 49, 50, 52, 53

[Lewitt 1992] R. M. Lewitt. *Alternatives to voxels for image representation in iterative reconstruction algorithms.* Phys. Med. and Biol., vol. 37, pages 705–716, 1992. 45, 55

[Li 2004] B. Li, G. B. Avinash, R. Uppaluri, J. W. Eberhard & B. E. H. Claus. *The impact of acquisition angular range on the z-resolution*

REFERENCES

of radiographic tomosynthesis. In International Congress Series, volume 1268, pages 13–18. Elsevier, 2004. 142

[Li 2006] B. Li, R. Saunders & R. Uppaluri. *Measurement of slice thickness and in-plane resolution on radiographic tomosynthesis system using modulation transfer function (MTF).* In Proceedings of SPIE, volume 6142, page 61425D, 2006. 142

[Li 2012] Y. Li, Y. Chen, Y. Hu, A. Oukili, L. Luo, W. Chen & Toumoulin C. *Strategy of computed tomography sinogram inpainting based on sinusoid-like curve decomposition and eigenvector-guided interpolation.* J. Opt. Soc. Am. Opt. Image. Sci. Vis., vol. 29, no. 1, pages 153–163, 2012. 100

[Littleton 1965] J. T. Littleton & F. S. Winter. *Linear laminagraphy. A simple geometric interpretation of its clinical limitations.* The American journal of roentgenology, radium therapy, and nuclear medicine, vol. 95, no. 4, pages 980–991, 1965. 23

[Lo 1988] S. C. B. Lo. *Strip and line path-integrals with a square pixel matrix - a unified theory for computational CT projections.* IEEE Trans. Med. Imag., vol. 7, no. 4, pages 355–363, 1988. 48

[Logan 1975] B. F. Logan & L. A. Shepp. *Optimal reconstruction of a function from its projections.* Duke Math. J., vol. 42, no. 4, pages 645–659, 1975. 101

[Long 2010a] Y. Long & J. A. Fessler. *3D forward and back-projection for X-ray CT using separable footprints with trapezoid functions.* In Proceedings of The First International Conference on Image Formation in X-Ray Computed Tomography, pages 216–219, 2010. 49

[Long 2010b] Y. Long, J. A. Fessler & James M. Balter. *3D forward and backprojection for X-ray CT using separable footprints.* IEEE Trans. Med. Imag., vol. 29, no. 11, pages 1839–1850, 2010. 49

[Long 2011] Y. Long, J. A. Fessler & James M. Balter. *3D forward and back-projection for X-ray CT using separable footprints.* In 11th International Meeting on Fully Three-Dimensional Image Reconstruction in Radiology and Nuclear Medicine, pages 146–149, 2011. 49

[Ludwig 2008] T. Ludwig J.and Mertelmeier, H. Kunze & W. Haerer. *A novel approach for filtered backprojection in tomosynthesis based on filter kernels determined by iterative reconstruction techniques.* In Elizabeth Krupinski, editeur, Digital Mammography, volume 5116

REFERENCES

of *Lecture Notes in Computer Science*, pages 612–620. Springer Berlin / Heidelberg, 2008. 25

[Machida 2010] H. Machida, T. Yuhara, T. Mori, E. Ueno, Y. Moribe & J. M. Sabol. *Optimizing parameters for flat-panel detector digital tomosynthesis.* Radiographics, vol. 30, no. 2, pages 549–562, 2010. 17, 143, 146, 149, 161

[Maidment 2006] A. Maidment, C. Ullberg, K. Lindman, L. Adelöw, J. Egerström, M. Eklund, T. Francke, U. Jordung, T. Kristoffersson, L. Lindqvist, D. Marchal, H. Olla, E. Penton, J. Rantanen, S. Solokov, Weber N. & H. Westerberg. *Evaluation of a photon-counting breast tomosynthesis imaging system.* In Proc. of SPIE, volume 6142, page 61420B, 2006. 146

[Maltz 2009] J. S. Maltz, F. Sprenger, J. Fuerst, A. Paidi, F. Fadler & A. R. Bani-Hashemi. *Fixed gantry tomosynthesis system for radiation therapy image guidance based on a multiple source x-ray tube with carbon nanotube cathodes.* Med. Phys., vol. 36, pages 1624–36, 2009. 153

[Marabini 1998] R. Marabini, G. T. Herman & J. M. Carazo. *3D reconstruction in electron microscopy using ART with smooth spherically symmetric volume elements (blobs).* Ultramicroscopy, vol. 72, no. 1-2, pages 53–65, 1998. 49

[Marchal 2011] C. Marchal. *Study of the Kepler's conjecture: the problem of the closest packing.* Mathematische Zeitschrift, vol. 267, no. 3-4, pages 737–765, 2011. 55

[Matej 1996] S. Matej & R. M. Lewitt. *Practical considerations for 3-D image reconstruction using spherically symmetric volume elements.* IEEE Trans. Med. Imag., vol. 15, no. 1, pages 68–78, 1996. 53, 54, 55

[Matsuo 1993] H. Matsuo, A. Iwata, I. Horiba & N. Suzumura. *Three-dimensional image reconstruction by digital tomosynthesis using inverse filtering.* IEEE Trans. Med. Imag., vol. 12, no. 2, pages 307–313, 1993. 25

[Mayer 1916] K. Mayer. *Radyologiczne rozpoznanie rozniczkowe chorób serca i aorty*, 1916. Patent, Gebethner and Co., Krakow. 19

[Merserea 1974] R. M. Merserea & A. V. Oppenhei. *Digital reconstruction of multidimensional signals from their projections.* Proc of IEEE, vol. 62, no. 10, pages 1319–1338, 1974. 37

REFERENCES

[Mertelmeier 2008] T. Mertelmeier, J. Ludwig, B. Zhao & W. Zhao. *Optimization of tomosynthesis acquisition parameters: angular range and number of projections*. In Digital Mammography, volume 5116 of *Lecture Notes in Computer Science*, pages 220–227. Springer, 2008. 143, 149

[Miller 1971] E. R. Miller, E. M. McCurry & B. Hruska. *An infinite number of laminagrams from a finite number of radiographs*. Radiology, vol. 98, no. 4, page 249Ű255, 1971. 20, 106

[Mueller 1997] K. Mueller, R. Yagel & J. F. Cornhill. *The weighted-distance scheme: a globally optimizing projection ordering method for ART*. IEEE Trans. Med. Imag., vol. 16, no. 2, pages 223–230, 1997. 88, 94

[Mueller 1998a] K. Mueller. *Fast and accurate three-dimensional reconstruction from cone-beam projection data using algebraic methods*. PhD thesis, Graduate School of The Ohio State University, 1998. 78, 79, 87, 88

[Mueller 1998b] K Mueller, R Yagel & JJ Wheller. *A fast and accurate projection algorithm for 3D cone-beam reconstruction with the algebraic reconstruction technique (ART)*. In Proc. SPIE, volume 3336, pages 1–9, 1998. 47

[Mumcuoglu 1994] E. U. Mumcuoglu, R. Leahy, S. R. Cherry & Z.Y. Zhou. *Fast gradient-based methods for bayesian reconstruction of transmission and emission PET images*. IEEE Trans. Med. Imag, vol. 13, no. 4, pages 687–701, 1994. 77

[Nikazad 2008] T. Nikazad. *Algebraic reconstruction methods*. PhD thesis, Department of Mathematics, Linköpings universitet, 2008. 79

[Niklason 1997] L. T. Niklason, B. T. Christian, L. E. Niklason, D. B. Kopans, D. E. Castleberry, B. H. Opsahl-Ong, C. E. Landberg, P. J. Slanetz, A. A. Giardino, R. Moore, D. Albagli, M. C. DeJule, P. F. Fitzgerald, D. F. Fobare, B. W. Giambattista, R. F. Kwasnick, Liu J., S.J. Lubowski, G. E. Possin, J. F. Richotte, C. Y. Wei & Wirth R. F. *Digital tomosynthesis in breast imaging*. Radiology, vol. 205, no. 2, pages 399–406, 1997. 12

[Nuyts 1997] J. Nuyts, P. Dupont & L. Mortelmans. *Iterative reconstruction of transmission sinograms with low signal to noise ratio*. In Karny Miroslav & Kevin Warwick, editeurs, Computer Intensive Methods in Control and Signal Processing, pages 237–248. Birkhäuser Boston, 1997. 81

REFERENCES

[Oehler 2007] M. Oehler & T. M. Buzug. *Statistical image reconstruction for inconsistent CT projection data*. Journal of Methods of Information in Medicine, vol. 3, pages 261–269, 2007. 125

[Ogawa 2010] K. Ogawa, R. P. Langlais, W. D. McDavid, M. Noujeim, K. Seki, T. Okano, T. Yamakawa & T. Sue. *Development of a new dental panoramic radiographic system based on a tomosynthesis method*. Dentomaxillofac Radiol., vol. 39, no. 1, pages 47–53, 2010. 12

[Papoulis 1962] A. Papoulis. *The fourier integral and its applications*. McGraw-Hill Co., New York, 1962. 37

[Park 2007] J. M. Park, E. A. J. R. Franken, M. Garg, L. L. Fajardo & L. T. Niklason. *Breast tomosynthesis: present considerations and future applications*. Radiographics, vol. 27, no. SI, pages S231–S240, 2007. 92nd Scientific Assembly and Annual Meeting of the Radiological-Society-of-North-America, Chicago, IL, 26.Nov-01.Dec, 2006. 12

[Peltonen 2010] S. Peltonen & U. Ruotsalainen. *Exact formulation of stackgram filters in sinogram domain*. In Nuclear Science Symposium Conference Record (NSS/MIC), IEEE, pages 3361–3365, 2010. 100

[Peters 1981] T. M. Peters. *Algorithms for fast back and re-projection in computed tomography*. IEEE Trans. Nucl. Sci., vol. 28, no. 4, pages 3641–3647, 1981. 47

[Petersson 1980] C. U. Petersson, P. Edholm, G. H. Granlund & H. E. Knutsson. *Ectomography - a new radiographic reconstruction method. 2. Computer-simulated experiments*. IEEE Trans. Biomed. Eng., vol. 27, no. 11, pages 649–655, 1980. 24

[Piccolomini 1999] E. L. Piccolomini & F. Zama. *The conjugate gradient regularization method in Computed Tomography problems*. Applied Mathematics and Computation, vol. 102, no. 1, pages 87–99, 1999. 77

[Pohl 1927] A. E. M. Pohl. *Method and apparatus for making Roentgen projections*, 1927. EU patent No 332496. 19

[Popescu 2004] L.M. Popescu & R.M. Lewitt. *Ray tracing through a grid of blobs*. In IEEE Nuclear Science Symposium Conference Record, volume 6, pages 3983–3986, 2004. 55

[Qian 2012] X. Qian, A. Tucker, E. Gidcumb, J. Shan, G. Yang, S. Calderon-Colon X .and Sultana, J. Lu, O. Zhou, D. Spronk, F. Sprenger, Y. Zhang, D. Kennedy, T. Farbizio & Z. Jing. *High resolution stationary digital breast tomosynthesis using distributed carbon*

REFERENCES

nanotube x-ray source array. Med. Phys., vol. 39, no. 4, pages 2090–2099, 2012. 153

[Qu 2009] C. Qu G.and Wang & M. Jiang. *Necessary and sufficient convergence conditions for algebraic image reconstruction algorithms.* IEEE Trans. Img. Proc., vol. 18, no. 2, pages 435–440, 2009. 79

[Quinto 1993] E. T. Quinto. *Singularities of the X-ray transform and limited data tomography in R(2) and R(3).* SIAM J. Math. Analysis, vol. 24, no. 5, pages 1215–1225, 1993. 35, 38, 154

[Quinto 2006] E. T. Quinto. *An antroduction to X-ray tomography and Radon transforms.* Proc. of Symposia in Appl. Math., vol. 63, pages 1–23, 2006. 35

[Quinto 2007] E. T. Quinto. *Local algorithms in exterior tomography.* J. Comput. Appl. Math., vol. 199, no. 1, pages 141–148, 2007. 35

[Quinto 2012] E. T Quinto. *Microlocal analysis in tomography.* http://www.helmholtz-muenchen.de/en/ibb/irma2012/IRMA_2012/Downloads_files/Todd_Quinto_Talk2.pdf, 2012. [Online; accessed 31-October-2012]. 33

[Radon 1917] J. Radon. *Über die Bestimmung von Funktionen durch ihre Integralwerte längs gewisser Mannigfaltigkeiten.* Ber. vor Sächs. Akad. Wiss., Leipzig Math. Phys, vol. 69, pages 262–277, 1917. 33, 34

[Radon 1986] J. Radon. *On the determination of functions from their integral values along certain manifolds.* IEEE Trans. Med. Imag., vol. 5, no. 4, pages 170–176, 1986. 34

[Reid 2011] C. B. Reid, M. M. Betcke, D. Chana & R. D. Speller. *The development of a pseudo-3D imaging system (tomosynthesis) for security screening of passenger baggage.* Nuclear Inst. and Methods in Physics Research A, vol. 652, no. 1, pages 108–111, 2011. 12

[Reiser 2007] I. Reiser, J. Bian, R. M. Nishikawa, E. Y. Sidky & X. Pan. *Comparison of reconstruction algorithms for digital breast tomosynthesis.* In Proc. of the 9th International Meeting on Fully Three-Dimensional Image Reconstruction in Radiology and Nuclear Medicine, pages 155–158, 2007. 39

[Reiser 2010] I. Reiser & R. M. Nishikawa. *Task-based assessment of breast tomosynthesis: effect of acquisition parameters and quantum noise.* Med. Phys., vol. 37, pages 1591–1600, 2010. 143, 149

[Ren 2006] B. Ren, T. Wu, A. Smith, C. Ruth, L. Niklason, Z. Jing & J. Stein. *The dependence of tomosynthesis imaging performance on the*

REFERENCES

                    number of scan projections. Digital Mammography, vol. 4046, pages 517–524, 2006. 146

[Richard 2010a] S. Richard & E. Samei. *Quantitative breast tomosynthesis: From detectability to estimability.* Med. Phys., vol. 37, no. 12, pages 6157–6165, 2010. 143

[Richard 2010b] S. Richard & E. Samei. *Quantitative imaging in breast tomosynthesis and CT: Comparison of detection and estimation task performance.* Med. Phys., vol. 37, no. 6, pages 2627–37, 2010. 143

[Roentgen 1895a] Wilhelm Conrad Roentgen. *Eine neue Art von Strahlen (II. Mittheilung)*, 1895. 1, 11, 19

[Roentgen 1895b] Wilhelm Conrad Roentgen. *Ueber eine neue Art von Strahlen (Vorläufige Mittheilung)*, 1895. 1, 11, 19

[Rorden, C. 2013] Rorden, C. *Web resource of Chris Rorden, a list of free DICOM viewers.* http://www.mccauslandcenter.sc.edu/mricro/dicom/index.html#links, 2013. [Online; accessed 20-February-2013]. 18

[Roy 1985] D. N. G. Roy, R. A. Kruger, B. Yih & P. Delrio. *Selective plane removal in limited angle tomographic imaging.* Med. Phys., vol. 12, no. 1, pages 65–70, 1985. 24

[Ruttimann 1984] U. E. Ruttimann, R. A. J. Groenhuis & Webber R. L. *Restoration of digital multiplane tomosynthesis by a constrained iteration method.* IEEE Trans. Med. Imag., vol. 3, no. 3, pages 141–148, 1984. 24

[Sauer 1993] K. Sauer & C. Bouman. *A local update strategy for iterative reconstruction from projections.* IEEE Trans. Sig. Proc., vol. 41, no. 2, pages 534–548, 1993. 77

[Schmitt 2012] K. Schmitt, H. Schoendube, K. Stierstorfer, J. Hornegger & F. Noo. *Analysis of bias induced by various forward projection models in iterative reconstruction.* In Proceedings of The Second International Conference on Image Formation in X-Ray Computed Tomography, pages 288–292, 2012. 45

[Sechopoulos 2009] I. Sechopoulos & C. Ghetti. *Optimization of the acquisition geometry in digital tomosynthesis of the breast.* Med. Phys., vol. 36, no. 4, pages 1199–1207, 2009. 142, 146

[Sechopoulos 2013a] I. Sechopoulos. *A review of breast tomosynthesis. Part I. The image acquisition process.* Med. Phys., vol. 40, no. 1, 2013. Article-Number 014301. 2

REFERENCES

[Sechopoulos 2013b] I. Sechopoulos. *A review of breast tomosynthesis. Part II. Image reconstruction, processing and analysis, and advanced applications.* Med. Phys., vol. 40, no. 1, 2013. Article-Number 014302. 2

[Segars 010] W. P. Segars, G. Sturgeon, S. Mendonca, J. Grimes & B. M. W. Tsui. *4D XCAT phantom for multimodality imaging research.* Med. Phys., vol. 37, no. 9, pages 4902–4915, 2010. 101

[Shepp 1982] L. A. Shepp & Y. Vardi. *Maximum Likelihood Reconstruction for Emission Tomography.* IEEE Trans. Med. Imag., vol. 1, no. 2, pages 113–122, 1982. 81

[Siddon 1985] R. L. Siddon. *Fast calculation of the exact radiological path for a 3-dimensional CT array.* Med. Phys., vol. 12, no. 2, pages 252–255, 1985. 48

[Smith 1985] B. D. Smith. *Image-reconstruction from cone-beam projections - necessary and sufficient conditions and reconstruction methods.* IEEE Trans. Med. Imag., vol. 4, no. 1, pages 14–25, 1985. 38, 154, 156

[Sone 1991a] S. Sone, T. Kasuga, F. Sakai, J. Aoki, I. Izuno, Y. Tanizaki, H. Shigeta & K. Shibata. *Development of a high-resolution digital tomosynthesis system and its clinical application.* Radiographics, vol. 11, no. 5, pages 807–822, 1991. 24

[Sone 1991b] S. Sone, T. Kasuga, F. Sakai, I. Izuno & M. Oguchi. *Digital image processing to remove blur from linear tomography of the lung.* Acta Radiol., vol. 32, no. 5, pages 421–425, 1991. 24

[Stevens 2001] G. M. Stevens, R. Fahrig & N. J. Pelc. *Filtered backprojection for modifying the impulse response of circular tomosynthesis.* Med. Phys., vol. 28, no. 3, pages 372–380, 2001. 25

[Stiel 1993] G. M. Stiel, L. S. G. Stiel, E. Klotz & C. A. Nienaber. *Digital flashing tomosynthesis - a promising technique for angiocardiographic screening.* IEEE Trans. Med. Imag., vol. 12, no. 2, pages 314–321, 1993. 24, 109

[Sun 2007] X. Sun, W. Land & R. Samala. *Deblurring of tomosynthesis images using 3D anisotropic diffusion filtering.* volume 6512, pages P1–P11, 2007. 24

[Svahn 2007] T. Svahn, M. Ruschin, B. Hemdal, I. Andersson, P. Timberg & A. Tingberg. *In-plane artifacts in breast tomosynthesis quantified with a novel contrast-detail phantom.* In Proc. SPIE, volume 6510, page 65104R, 2007. 39

REFERENCES

[Tabei 1992] M. Tabei & M. Ueda. *Backprojection by upsampled Fourier series expansion and interpolated FFT.* IEEE Trans. Imag. Proc., vol. 1, no. 1, pages 77–87, 1992. 48

[Thomson 1896] E. Thomson. *Stereoscopic Roentgen pictures.* Electr. Eng., vol. 21, page 256, 1896. 19

[Tingberg 2010] A. Tingberg. *X-ray tomosynthesis: a review of its use for breast and chest imaging.* Radiat. Prot. Dosim., vol. 139, no. 1-3, pages 100–107, 2010. 12

[Toft 1996] P. Toft. *The Radon transform - theory and implementation.* PhD thesis, Department of Mathematical Modelling, Technical University of Denmark, June 1996. [Online; accessed 21-January-2013]. 46, 79

[Tuy 1983] K. H. Tuy. *An inversion formula for cone-beam reconstruction.* SIAM J. Appl. Math., vol. 45, no. 3, pages 456–552, 1983. 38, 154, 156

[Upton 1992] A. C. Upton. *The first hundred years of radiation research: what have they taught us?* Environ. Res., vol. 59, no. 1, pages 36–48, 1992. 19

[Vallebona 1932] A. Vallebona. *Una modalità di tecnica per la dissociazione radiografica delle ombre applicata allo studio del cranio.* Radiol. Med., vol. 17, pages 1090–1097, 1932. 19

[van de Sompel 2007] D. van de Sompel. *Development and validation of reconstruction algorithms for digital breast tomosynthesis.* PhD thesis, Medical Vision Laboratory, Department of Engineering Science, University of Oxford, September 2007. 79

[van der Stelt 1986] P. F. van der Stelt, U. E. Ruttimann & R. L. Webber. *Enhancement of tomosynthetic images in dental radiology.* Journal Of Dental Research, vol. 65, no. 7, pages 967–973, 1986. 24

[van Dijke 1992] M. C. A. van Dijke. *Iterative methods in image reconstruction.* PhD thesis, State University Utrecht, The Netherlands, November 1992. Publication 557. 88

[Van Tiggelen 2002] R. Van Tiggelen. *In search for the third dimension: from radiostereoscopy to three-dimensional imaging.* Belgian Journal of Radiology(JBR-BTR), vol. 85, no. 5, pages 266–270, 2002. 20

[Wang 2004] W. L. Wang, W. Hawkins & D. Gagnon. *3D RBI-EM reconstruction with spherically-symmetric basis function for SPECT rotating slat collimator.* Phys. Med. Biol., vol. 49, no. 11, pages 2273–2292, 2004. 49

REFERENCES

[Wang 2007] J. Wang & Y. Zheng. *On the convergence of generalized simultaneous iterative reconstruction algorithms.* IEEE Trans. Img. Proc., vol. 16, no. 1, pages 1–6, JAN 2007. 79

[Wang 2010] H. Wang, L. Desbat & S. Legoupil. *2D X-ray CT reconstruction based on TV minimization and blob representation.* In The First International Conferenceon Image Formation in X-ray Computed Tomography, pages 252–255, 2010. 57

[webDICOMLOOKUP 2013] webDICOMLOOKUP. *dicomlookup.* http://dicomlookup.com/, 2013. [Online; accessed 20-February-2013]. 18

[webFDA 2013] webFDA. *Web resource of U.S. Food and Drug Administration, Selenia Dimensions 3D System - P080003.* http://www.fda.gov/MedicalDevices/ProductsandMedicalProcedures/DeviceApprovalsandClearances/Recently-ApprovedDevices/ucm246400.htm, 2013. [Online; accessed 20-February-2013]. 21

[webMATLAB 2013] webMATLAB. *Web resource of MATLAB® file exchange.* http://www.mathworks.com/matlabcentral/fileexchange/, 2013. [Online; accessed 20-February-2013]. 167

[webNEMA 2013] webNEMA. *Web resource of the National Electrical Manufacturers Association.* http://medical.nema.org, 2013. [Online; accessed 20-February-2013]. 17

[webPUBMED 2013] webPUBMED. *Web resource of the US National Library of Medicine National Institutes of Health.* http://www.ncbi.nlm.nih.gov/pubmed/, 2013. [Online; accessed 20-February-2013]. 21, 22

[Wheatstone 1838] Charles Wheatstone. *Contributions to the physiology of vision. Part the first. On some remarkable, and hitherto unobserved, phenomena of binocular vision.* Phil. Trans. R. Soc. London, vol. 128, pages 371–394, 1838. 18

[Wu 2004] T. Wu, R. H. Moore, E. A. Rafferty & D. B. Kopans. *A comparison of reconstruction algorithms for breast tomosynthesis.* Med. Phys., vol. 31, no. 9, pages 2636–2647, 2004. 25

[Wu 2006] T. Wu, R. H. Moore & D. B. Kopans. *Voting strategy for artifact reduction in digital breast tomosynthesis.* Med. Phys., vol. 33, no. 7, pages 2461–2471, 2006. 24, 109

[Wu 2008] G. Wu, J. Mainprize & M. Martin J. Yaffe. *Characterization of projection ordering in iterative reconstruction methods for breast tomosynthesis.* In IWDM, volume LNCS 5116, pages 601–605. Springer Berlin Heidelberg, 2008. 87

REFERENCES

[Wu 2010] G. Wu, J. G. Mainprize & M. J. Yaffe. *Breast tomosynthesis reconstructionusing a grid of blobs with projection matrices*. In Digital Mammography, Lecture Notes in Computer Science, pages 243–250. Springer Berlin Heidelberg, 2010. 49

[Wu 2011] M. Wu & J.A. Fessler. *GPU acceleration of 3D forward and backward projection using separable footprints for X-ray CT image reconstruction*. In 3rd Workshop on High Performance Image Reconstruction, pages 53–56, 2011. 49

[Xu 2006] Fang Xu & K. Mueller. *A comparative study of popular interpolation and integration methods for use in computed tomography*. In Biomedical Imaging: Nano to Macro, 2006. 3rd IEEE International Symposium on, pages 1252–1255, 2006. 48

[Yendiki 2004] A. Yendiki & J. A. Fessler. *A comparison of rotation- and blob-based system models for 3D SPECT with depth-dependent detector response*. Phys. Med. Biol., vol. 49, no. 11, pages 2157–2168, 2004. 49

[Zamyatin 2007] A. A. Zamyatin & S Nakanishi. *Extension of the reconstruction field of view and truncation correction sing sinogram decomposition*. Med. Phys., vol. 34, no. 5, pages 1593–1604, 2007. 100

[Zbijewski 2006] W. Zbijewski & F. J. Beekman. *Comparison of methods for suppressing edge and aliasing artefacts in iterative x-ray CT reconstruction*. Phys. Med. Biol., vol. 51, no. 7, pages 1877–1889, 2006. 49

[Zeng 2000] G. S. L. Zeng & G. T. Gullberg. *Unmatched projector/backprojector pairs in an iterative reconstruction algorithm*. IEEE Trans. Med. Imag., vol. 19, no. 5, pages 548–555, 2000. 48

[Zhang 2006a] B. Zhang & G. L. Zeng. *An immediate after-backprojection filtering method with blob-shaped window functions for voxel-based iterative reconstruction*. Phys. Med. Biol., vol. 51, no. 22, pages 5825–5842, 2006. 49

[Zhang 2006b] Y. Zhang, H. Chan, B. Sahiner, J. Wei, M. M. Goodsitt, L. M. Hadjiiski, J. Ge & C. Zhou. *A comparative study of limited-angle cone-beam reconstruction methods for breast tomosynthesis*. Med. Phys., vol. 33, no. 10, pages 3781–3795, 2006. 25, 82

[Zhang 2010] J. Zhang & C. Yu. *A novel solid-angle tomosynthesis (SAT) scanning scheme*. Med. Phys., vol. 37, no. 8, pages 4186–4192, 2010. 152, 153

REFERENCES

[Zhao 2003] H. Zhao & A. J. Reader. *Fast ray-tracing technique to calculate line integral paths in voxel arrays*. In Nuclear Science Symposium Conference Record, IEEE, volume 4, pages 2808–2812, 2003. 48

[Zhao 2008] B. Zhao & W. Zhao. *Three-dimensional linear system analysis for breast tomosynthesis*. Med. Phys., vol. 35, pages 5219–5232, 2008. 143

[Zhou 2007] J. Zhou, B. Zhao & W. Zhao. *A computer simulation platform for the optimization of a breast tomosynthesis system*. Med. Phys., vol. 34, pages 1098–1109, 2007. 142, 146

[Ziedses des Plantes 1932] B. G. Ziedses des Plantes. *Eine neue Methode zur Differenzierung in der Röntgenographie (Planigraphie)*. Acta Radiol., vol. 13, pages 182–192, 1932. 16, 19, 106

[Ziegler 2006] T. Ziegler A.and Koehler, T. Nielsen & R. Proksa. *Efficient projection and backprojection scheme for spherically symmetric basis functions in divergent beam geometry*. Med. Phys., vol. 33, no. 12, pages 4653–4663, 2006. 57

Aktuelle Forschung Medizintechnik

Herausgeber:

Prof. Dr. Thorsten M. Buzug

Institut für Medizintechnik, Universität zu Lübeck

Editorial Board:
Prof. Dr. Olaf Dössel, Karlsruhe Institute for Technology; Prof. Dr. Heinz Handels, Universität zu Lübeck; Prof. Dr.-Ing. Joachim Hornegger, Universität Erlangen-Nürnberg; Prof. Dr. Marc Kachelrieß, German Cancer Research Center (DKFZ), Heidelberg; Prof. Dr. Edmund Koch, TU Dresden; Prof. Dr.-Ing. Tim C. Lüth, TU München; Prof. Dr. Dietrich Paulus, Universität Koblenz-Landau; Prof. Dr. Bernhard Preim, Universität Magdeburg; Prof. Dr.-Ing. Georg Schmitz, Universität Bochum.

Themen
Werke aus folgenden Themengebieten werden gerne in die Reihe aufgenommen: Biomedizinische Mikro- und Nanosysteme, Elektromedizin, biomedizinische Mess- und Sensortechnik, Monitoring, Lasertechnik, Robotik, minimalinvasive Chirurgie, integrierte OP-Systeme, bildgebende Verfahren, digitale Bildverarbeitung und Visualisierung, Kommunikations- und Informationssysteme, Telemedizin, eHealth und wissensbasierte Systeme, Biosignalverarbeitung, Modellierung und Simulation, Biomechanik, aktive und passive Implantate, Tissue Engineering, Neuroprothetik, Dosimetrie, Strahlenschutz, Strahlentherapie.

Autorinnen und Autoren
Autoren der Reihe sind in der Regel junge Promovierte und Habilitierte, die exzellente Abschlussarbeiten verfasst haben.

Leserschaft
Die Reihe wendet sich einerseits an Studierende, Promovenden und Habilitanden aus den Bereichen Medizintechnik, Medizinische Ingenieurwissenschaft, Medizinische Physik, Medizinische Informatik oder ähnlicher Richtungen. Andererseits stellt die Reihe aktuelle Arbeiten aus einem sich schnell entwickelnden Feld dar, so dass auch Wissenschaftlerinnen und Wissenschaftler sowie Entwicklerinnen und Entwickler an Universitäten, in außeruniversitären Forschungseinrichtungen und der Industrie von den ausgewählten Arbeiten in innovativen Gebieten der Medizintechnik profitieren werden.

Begutachtungsprozess
Die Qualitätssicherung erfolgt in drei Schritten. Zunächst werden nur Arbeiten angenommen die mindestens magna cum laude bewertet sind. Im zweiten Schritt wird ein Mitglied des Editorial Boards die Annahme oder Ablehnung des Werkes empfehlen. Im letzten Schritt wird der Reihenherausgeber über die Annahme oder Ablehnung entscheiden sowie Änderungen in der Druckfassung empfehlen. Die Koordination übernimmt der Reihenherausgeber.

Kontakt
Prof. Dr. Thorsten M. Buzug
Institut für Medizintechnik   Tel.:   +49 (0) 451 / 500-5400
Universität zu Lübeck   Fax: +49 (0) 451 / 500-5403
Ratzeburger Allee 160   E-Mail: buzug@imt.uni-luebeck.de
23538 Lübeck, Germany   Web: http://www.imt.uni-luebeck.de

Stand: Januar 2014. Änderungen vorbehalten.
Erhältlich im Buchhandel oder beim Verlag.

Abraham-Lincoln-Straße 46
D-65189 Wiesbaden
Tel. +49 (0)6221. 345 - 4301
www.springer-vieweg.de

MIX
Papier aus verantwortungsvollen Quellen
Paper from responsible sources
FSC® C105338

If you have any concerns about our products,
you can contact us on
ProductSafety@springernature.com

In case Publisher is established outside the EU,
the EU authorized representative is:
Springer Nature Customer Service Center GmbH
Europaplatz 3, 69115 Heidelberg, Germany

Printed by Libri Plureos GmbH
in Hamburg, Germany